The Pentagon Labyrinth

10 Short Essays to Help You Through It

From 10 Pentagon Insiders, Retired Military Officers and Specialists
With Over 400 Years of Defense Experience

Edited by Winslow T. Wheeler

Center for Defense Information
World Security Institute
February 2011

About the Authors

Thomas Christie began his career in the Department of Defense and related positions in 1955. He retired from the Pentagon in February 2005 after four years as Director of Operational Test & Evaluation. There he was responsible for policy and procedures for testing weapon systems and for providing independent evaluations of the test results to both the defense secretary and Congress. He earlier served as director of the Operational Evaluation Division at the Institute for Defense Analyses, where he was also involved in DOD weapons testing. Between 1985 and 1989, he was director of program integration in the Office of the Secretary of Defense, responsible for developing procedures for managing the defense acquisition system. Prior to that, he had served in two separate positions under the assistant secretary of defense (Program Analysis and Evaluation): director of tactical air division and deputy assistant secretary of defense for General Purpose Programs.

Andrew Cockburn is a writer and documentary filmmaker resident in Washington, D.C. He has covered defense and national security issues for over 30 years. He has authored several books, including *The Threat: Inside the Soviet Military Machine* (1982), *Out of the Ashes, The Resurrection of Saddam Hussein* (1999) and *Rumsfeld: His Rise, Fall and Catastrophic Legacy* (2007).

Bruce I. Gudmundsson served in the Marine Corps Reserve for 20 years, joining as a private in 1977 and retiring as a major in 1997. The author of seven books and several hundred articles, he is a historian who specializes in the internal workings of military forces (their structure, training, doctrine and culture), as well as the way that these things influence their ability to adapt to changing circumstances.

Col. Chet Richards (U.S. Air Force, ret.) is a consultant and writer with J. Addams & Partners in Atlanta. He is the author of *If We Can Keep It: A National Security Manifesto for the Next Administration* (2007), *Certain to Win: The Strategy of John Boyd Applied to Business* (2004) and other publications on Third and Fourth Generation Warfare. He holds a doctorate in mathematics and is adjunct professor of strategy and quantitative analysis at Kennesaw State University.

Franklin C. Spinney retired from the Defense Department in 2003 after a military-civilian career spanning 33 years. The latter 26 of those years were as a staff analyst in the Office of the Secretary of Defense. During this period, he appeared as a witness in numerous congressional hearings before the Budget, Armed Services, and Government Affairs or Reform and Oversight committees of the U.S. House and Senate. He is author of *Defense Facts of Life: The Plans/Reality Mismatch* (1985). His op-eds and essays have appeared in the *The*

Wall Street Journal, The Washington Post, Los Angeles Times, Challenge, CounterPunch, Proceedings Magazine of the U.S Naval Institute and the *Marine Corps Gazette*, among other places. His sharply critical analysis of the Reagan defense program landed him on the cover of *Time Magazine* (March 7, 1983), based on a hearing at which the senior Pentagon management witness promised all Pentagon budgeting and programming problems were being effectively dealt with. In 2003, he held an hour long "exit interview" with Bill Moyers on Moyers' PBS show *NOW;* the basic message was that 20 years later, none of the problems had been addressed, let alone solved.

Pierre Sprey consulted for Grumman Aircraft's research department from 1958 to 1965, then joined Secretary of Defense Robert McNamara's "Whiz Kids" in the Pentagon. There, in 1967, he met the Air Force's brilliant and original tactician, Col. John Boyd and quickly became a disciple and collaborator of Boyd's. Together with another innovative fighter pilot, Col. Everest Riccioni (U.S. Air Force), they started and carried out the concept design of the F-16 air-to-air fighter, then brought the program to fruition through five years of continuous bureaucratic guerilla warfare. More or less simultaneously, Sprey also headed up the technical side of the Air Force's concept design team for the A-10 close support fighter. Then, against even steeper opposition than the F-16 faced, he helped implement the A-10's innovative live-fire, prototype fly-off competition and subsequent production. Sprey left the Pentagon in 1971 but continued to consult actively on the F-16, the A-10, tanks and anti-tank weapons, and realistic operational/live-fire testing of major weapons. At the same time, he became a principal in two consulting firms; the first doing environmental research and analysis, the second consulting on international defense planning and weapons analysis. During this period, Sprey continued the seminal work of Col. Richard Hallock (U.S. Army/Airborne) in founding the field of combat history/combat data-based cost effectiveness analysis for air and ground weapons. During the late 1970s, Colonel Boyd and Sprey, together with a small, dedicated group of Pentagon and congressional insiders, started the military reform movement. Attracting considerable attention from young officers, journalists and congressmen, the movement led to establishment of the Congressional Military Reform Caucus and to passage of several military reform bills in the early '80s. Sprey continues to work with reform-minded foundations and journalists. Numerous articles, books and theses have described the work of Colonel Boyd and Sprey on the F-16, A-10 and military reform. These include Robert Coram's "Boyd: The Fighter Pilot Who Changed the Art of War" (2002) and James Fallows' "National Defense" (1981).

Winslow T. Wheeler is the director of the Straus Military Reform Project at the Center for Defense Information in Washington, D.C. He has authored two books: *The Wastrels of Defense* (2004) about Congress and national security, and *Military Reform* (2007). He is the editor of *America's Defense Meltdown* (2008) and of this handbook. From 1971 to 2002, Wheeler worked on national security issues for members of the U.S. Senate from both political parties and for the U.S. Government Accountability Office (GAO). In 2002, he was forced to resign his position with the Republican staff of the Senate Budget Committee because of senators' objections to an essay he wrote, "Mr. Smith Is Dead: No One Stands in the Way as Congress Lards Post-September 11 Defense Bills with Pork."

George C. Wilson has had an upfront seat from which to study the military-industrial-political-intelligence complex for a half century. After five years with *Aviation Week & Space Technology* magazine, which gave him an insight into the inner workings of the complex, he became the chief military correspondent for *The Washington Post*. He did combat reporting for *The Post* in South Vietnam, the Middle East and Panama. He is the author of six books, including the best seller *Supercarrier,* based on his seven-and-a- half month deployment on the aircraft carrier *John F. Kennedy,* during which time he flew in every plane on her deck. Wilson is a Navy veteran, graduate of Bucknell University and winner of several journalism awards. He resides in Arlington, Va.

G.I. Wilson (U.S. Marine Corps, ret.) has deployed on multiple combat tours with the Marines; he holds a bachelor's degree from the State University of New York - Albany and a master's degree from Webster University. He teaches criminal justice at the college level and is a graduate student in forensic psychology currently doing an internship with University of California, San Diego's Department of Psychiatry.

Dedication and Acknowledgements

The Pentagon Labyrinth is dedicated to three people.

To Philip A. Straus Jr., without whom *The Pentagon Labyrinth* would not have existed. In addition, without his support the work of the Straus Military Reform Project of the Center of Defense Information would also not have existed. His generosity also extends to the spirit of his support, which has never included a demand or even a request but only the gentle guidance that has made him a pleasure to work with. If there were more like him in Washington, the national capital would not be the howling cacophony that exists today.

To Col. John R. Boyd, U.S. Air Force, ret., whose spirit infuses the work of every author in this handbook; I know of no higher praise to offer to the others involved in the writing of *The Pentagon Labyrinth,* most—but not all of them listed as the authors. (To find out more about Boyd, see the first section of the Suggested Readings of this handbook.)

To Maj. Donald E. Vandergriff, U.S. Army, ret., whom all of us have had the honor to work with. Don embodies the message of this handbook by doing in his career, as an officer and as a civilian, the tireless and thankless work the institution he so dearly loves, the United States Army, needs more than it knows. Don has chosen for his life an exemplar path for the ethical, educational behavior that this handbook seeks to encourage. (Most of Don's public books and monographs are listed in the Suggested Readings section of this handbook.)

I also wish to thank the staff of the World Security Institute for their spirited support, diligent work and skillful professionalism in the completion of this handbook. To be noted are Bruce Blair, Drew Portocarrero, Ollie Harrison-Little and—especially—Goran Hinrichs, and his able assistant Janice Romzek.

Winslow T. Wheeler, Editor

Table of Contents

Preface

This handbook aims to help newcomers as well as seasoned observers learn how to grapple with the problems of national defense, using insights our authors have gleaned in the course of their more than 400 years of combined experience.

Each year, people are hired to address defense issues in the Pentagon, on Capitol Hill, in think tanks and throughout the media. Some of them will have experience in the armed forces; some have studied national security in universities, some have worked in the Pentagon or the defense industry. Many of them might consider a handbook for defense "newbies" to be beneath them, but few of them will have had the depth of experience across all the disciplines represented by the authors of this book: decades in military service, intelligence, weapons design, Pentagon defense management and analysis, weapons testing, journalism, military history and congressional staff work.

The Pentagon Labyrinth is intended to benefit defense professionals in the early stages of their career, but it very probably has some worthwhile lessons for people who have been working in national security for a long time. It is not just that the conventional defense wisdom (resting on clichés such as "American military technology gives us the winning edge") is so often misinformed. It is also that experienced journalists, senior congressional staffers and seasoned Pentagon officials too often take in and pass on these bromides without thinking about their implications, intended or unintended. Examples abound:

- How many times does one read articles stating the cost of a weapon—the F-35 is a contemporary example—as described by a hired consultant for a manufacturer or an advocate from inside the Pentagon? That price tag is published as if it were authoritative; there's not a hint that more objective sources would cite a very different figure. The handbook's essay on journalism ("Penetrating the Pentagon" by George Wilson), as well as the one on costs, might help journalists reporting on weapons serve their readers better, and those essays might help readers more effectively identify the journalists they may want to read more, or less, from in the future.

- It is not just conventional wisdom but biblical text that the F-22 is a world class fighter aircraft; almost no one believes anything else. The ninth essay in this handbook ("Evaluating Weapons: Sorting the Good from the Bad" by Pierre Sprey) can start the reader on an adventure that leads to a very different conclusion.

- Herds of analysts, each with decades of experience inside the Washington Beltway, read with great seriousness the Pentagon's periodic "Quadrennial Defense Review" and opine on its contents—without appreciating that it is fundamentally a sham analysis of the Pentagon's problems. The first essay here ("Why Is This Handbook Necessary?" by Chuck Spinney) will explain.

- Seasoned staffers on Capitol Hill have taken offense at the suggestion that senior Pentagon civilians and high ranking military officers would lie to them. Yet the Constitution's system of checks and balances and the separation of powers in our federal government were conceived on just that premise: that interested factions in the Pentagon bureaucracy could—and do—go to great lengths not only to mask what is going on inside DOD but actively to present an alternate picture. The essay "Congressional Oversight: Willing and Able or Willing to Enable?" seeks to explain further.

The authors respectfully submit that even those who consider themselves expert in Pentagon matters can find something useful to learn in this handbook. Indeed, all of us who are the authors here have—simply by reading each other's essays.

The format of *The Pentagon Labyrinth* may be a little different from what most readers are accustomed to. Each section is a brief essay, not a chapter. We have tried to make these short and readable, rather than dry academic exercises. The footnotes are at the bottom of each page, not only to show sources but also to provide explanations and some additional, thought-provoking references to allow the interested reader to probe more deeply. The footnote links in electronic copies of *The Pentagon Labyrinth* should come to readers as active links. For the hard copy we have tried to make the footnoted URLs easy to transcribe.

The handbook ends with a list of suggested readings, contributed by the authors. These readings are what we believe to be unusually informative documents that provide valuable further insights into the defense problems introduced in each essay. Many of the references are hard to find elsewhere; some have never been published before, even on the Internet; a few of them are of historic significance—even if they have been hard to impossible to find up to now.

We have also created two Web sites for the entire text of this handbook and the informational materials. Items not previously available on the Internet were scanned to be electronically available for this handbook. These include selections of the works of Chuck Spinney and Pierre Sprey that are not otherwise accessible, a classic article by Dr. Thomas Amlie on the vulnerability of radar, unpublished Pentagon reports and other hard-to-find, invaluable materials. Download any of the essays or other materials at the Web sites for the Straus Military Reform Project of the Center for Defense information (at http://www.cdi.org/program/index.cfm?programid=37, or www.cdi.org/smrp)

and for the Project on Government Oversight (POGO) (at http://dnipogo.org/labyrinth/).

We expressly encourage you to download the handbook. The Center for Defense Information copyrighted the material for technical reasons, but the copyright will not be enforced against anyone who downloads the files of *The Pentagon Labyrinth* and who makes our text available without charge to anyone else. In fact, we encourage you to circulate the handbook liberally, or even to create your own Web page for it.

As you read this handbook, you will surely come across passages you will disagree with. If you find yourself saying "That can't be true!"—or something pithier—we encourage you to delve into the sources for that passage. If the available sources don't answer your doubts, contact the author and ask him to explain further or to provide you with more material. The email address of each author is listed on the first page of the section titled "Suggested Contacts, Readings and Web Sites." This was done specifically because our authors are seriously committed to the aim of this handbook: helping the reader think more clearly about defense problems.

The handbook follows a logical order. We start with Chuck Spinney's "Why Is This Handbook Necessary?" to address the underlying moral, intellectual and physical decay that besets our armed forces. The next four essays address how to approach "people" issues, overwhelmingly the most important ingredient of any successful military force. Col. Chet Richards' sixth essay addresses the next most important ingredient: "ideas" and the deficiencies in our strategic thinking. The last four essays address how to tackle our all-too-painful physical problems: money and budgets, weapons, testing and the buying of weapons. On the other hand, the handbook can be read in any order that interests you; each essay is self-standing.

Though each essay is also short, we hope they stimulate a continuing stream of new insights as you dig into the materials provided and use them to expand your contacts with the informed and ethical people we hope you will find based on your experience with *The Pentagon Labyrinth*.

Winslow T. Wheeler, Editor

Essay 1

"Why Is This Handbook Necessary?"

by Franklin C. Spinney

People say the Pentagon does not have a strategy. They are wrong.

The Pentagon does have a strategy; it is: 'Don't interrupt the money flow, add to it'

Col. John R. Boyd (U.S. Air Force, ret.)
Fighter Pilot, Tactician, Strategist,
Conceptual Designer, Reformer

Today, 20 years after the end of the Cold War and the disappearance of the Soviet Union, the United States spends more on defense than at any time since the end of World War II. This is true even if one removes the cumulative effects of 65 years of inflation from the current defense budget. Yet, notwithstanding the absence of a nuclear-armed superpower to threaten our existence, this gigantic defense budget is not producing a greater sense of security for most Americans.

Indeed, we have become a fearful nation, a bunkered nation, bogged down in never ending wars abroad accompanied by shrinking civil liberties at home. We now spend almost as much on defense as the rest of the world combined, yet the sinews of our supporting economy, particularly the all-important manufacturing sector, are weakening at an alarming rate, threatening the existence of the high-income, middle-class consumer society we built after World War II.

President Obama promised change, but he is under intense pressure to increase the defense budget even further, in part because he is continuing his predecessors' war-centric foreign policy. At the same time, he is being pressured to reduce the rapidly increasing federal deficit, caused in part by the rising defense budget, but also by an ill-advised bank bailout and the cyclical effects of the worst recession since the end of World War II. Moreover, the president initially promised to place the Pentagon off limits, while he sought reductions in discretionary spending for civilian programs and only reluctantly put defense spending "on the table" when he convened a bipartisan panel to seek a comprehensive path to a balanced budget. Lurking in the background, hanging over the American people like a guillotine, lies the menacing possibility of cutting Social Security and Medicare. In short, Obama may have promised change, but he is continuing the establishment's business-as-usual practices, including the grotesque diversion of scarce resources to a bloated defense budget that is leading the United States into ruin. Whether or not Obama's defense policy is a question of his free will is quite beside the point.

The salient question is: How did the American political system maneuver itself into such a destructive straitjacket?

This handbook is intended to provide readers - particularly students of defense, young military professionals, new Capitol Hill staff and concerned citizens - with the tools to understand the Pentagon's contribution to this mess and what might be needed to clean it up. We will speak to not just the insatiable demands for ever larger defense budgets, but also the directly resulting damage to America's defenses and to the integrity of its politics. And, most importantly, we hope to provoke thought on reversing that pervasive damage.

Follow the Money Trail

One source of the pressure for more defense spending is that our two relatively small wars in Iraq and Afghanistan, both much smaller in scale than the Korean or Vietnam wars, have stretched our military to the breaking point.[1] This is not to say that the day-to-day combat our troops face is any less grueling. On the contrary, our troops are stressed out, exhausted and many are traumatized by the intensity of their experiences - all worsened by the endless troop rotations caused by a military manpower base that is too small to sustain even these small wars. Moreover, despite the doubling of the defense budget since 1998, equipment and weapons are being worn out and not replaced, something that did not happen in either Vietnam or Korea.[2] The inventory is aging rapidly and modernization is going down the tubes because the new weapons the military

---

[1] These wars are small in terms of scale and tempo of operations. Bear in mind that the Korean and Vietnam wars took place against a backdrop of Cold War commitments. Today, the United States is spending more than we did in 1969 when we had 550,000 troops in Vietnam. But the Cold War meant that we also maintained hundreds of thousands of troops in Western Europe and East Asia, a huge rotation base at home to support these forward deployments, a large Navy fleet of 579 ships (compared to 287 today) to control the seas, and thousands of nuclear weapons on hair-trigger alert in airborne bombers, missile silos and submarines. Nevertheless, according to a report issued by the Congressional Research Service, the cumulative costs of the fighting in Afghanistan and Iraq have made the response to Sept. 11 the second most expensive war in U.S. history, exceeded only by World War II ("Cost of Major US Wars," CRS RS 22926, June 29, 2010; find it at http://www.fas.org/sgp/crs/natsec/RS22926.pdf).

[2] For example, during the Vietnam War, the Air Force modernized its inventory of F-100s and F-105s with considerably more expensive F-4s, A-7s and F-111s.

services choose to buy are many times more expensive than their predecessors. Therefore, the Pentagon cannot possibly buy enough new weapons to replace existing weapons one for one - even with a defense budget that has almost doubled since 1998.[3]

This current-war problem is a symptom of a deeper, more subtle web of intractable defense pathologies. These pathologies flow out of military-bureaucratic belief systems and distorted financial incentives that evolved slowly over the 40 years encompassing the Cold War. These pathologies and belief systems slowly insinuated themselves deeply and almost invisibly into a domestic political economy that nurtures financial-political factions of the Military - Industrial - Congressional Complex (MICC). The result is a voracious appetite for money that is sustained by a self-serving flood of ideological propaganda, cloaked by a stifling climate of excessive secrecy. President Eisenhower warned us to guard against the corrosive danger of exactly this in his 1961 farewell address.[4] He was ignored, and today, 50 years later, the domestic political imperative to steadily increase the money flowing into the MICC reaches into every corner of our society. It distorts and debases our economy, our politics, our universities and schools, our media, our think tanks and our research labs, just as Eisenhower predicted it would. Even without the Iraq and Afghanistan wars to hype the money flow, Mr. Obama could not have escaped massive pressures to increase defense spending.

In retrospect, it is clear that the Cold War served as a domestic political engine to keep the money flowing into the MICC. Many believed, erroneously as it turned out, the end of the Cold War would produce a "peace dividend" that would shut down the MICC and return the United States to being a normal country engaged primarily in peaceful business, not war.[5] However, by 1991, a true peace dividend would have collapsed the defense industry and its powerful political dependents. To survive and flourish, the factions of the MICC badly

---

[3] See "National Defense Budget Estimates for Fiscal Year 2011," Table 6-8: $708 billion (amended to include the requested funding for war spending in 2011) compared to $370 billion in Fiscal Year 1998 (converted to constant FY 2011 dollars) represents an increase of 91 percent, if one uses DOD's official inflation indices, available in Chapter 5 of the same National Defense Budget Estimates for FY 2011. Warning: DOD's inflation indices are self-serving and can exaggerate the effects of past inflation, thus reducing the apparent increase in today's budgets. Find "National Defense Budget Estimates for 2011" (also known as the "Green Book") at the DOD Comptroller's Web site at http://comptroller.defense.gov/Budget2011.html.

[4] Find a copy and video of this address at http://www.americanrhetoric.com/speeches/dwightdeisenhowerfarewell.html.

[5] A pamphlet I authored, "Defense Power Games" (Fund for Constitutional Government, 1990; download available at http://pogoarchives.org/labyrinth/01/09.pdf), explains why the belief in a peace dividend was fallacious; however, I failed to predict the MICC's dangerous mutation.

needed to evolve a subtle, pervasive shift in strategy, a subliminal mutation in the MICC's political DNA. It is now clear that this mutation has taken a frightening form: namely, the need to foment an enduring voter-terrifying threat and unending small wars to justify the money flow needed for the MICC's survival.

Without that never-ending succession of little wars (Somalia, Bosnia, Kosovo, the first and second Gulf wars, Afghanistan, Yemen, Pakistan, the war on terror, etc.) keeping the political system lathered up, the MICC's political-economic house of cards would collapse. A little reflection reveals that this mutation actually started in earnest as early as 1990, when Saddam Hussein invaded Kuwait. Clearly Sept. 11 did not create this mutation, but it certainly proved a windfall for expanding the scale and cost of our small wars.

Continuing small wars (or the threat thereof) are essential for the corporate component of the MICC; these companies have no alternative means to survive. Although they now make up a very substantial part of America's much diminished industrial base, they cannot convert to civilian production. Many of them tried and failed; they simply do not have the marketing, managing, engineering and manufacturing skills to compete successfully in global commercial markets. In the prophetic words of William Anders, CEO of General Dynamics in 1991, "... most [weapons manufacturers] don't bring a competitive advantage to non-defense business," and "Frankly, sword makers don't make good and affordable plowshares."[6]

---

[6] "Rationalizing America's Defense Industry: Renewing Investor Support for the Defense Industrial Base and Safeguarding National Security," Keynote Address by William Anders, Chairman and Chief Executive Officer, General Dynamics Corporation, presented to *Defense Week*, 12th Annual Conference, October 30, 1991, 13. Anders' intent was to explain why General Dynamics was not going to diversify its business into the non-defense sector, given the end of the Cold War. He rationalized a takeover strategy to increase market share in a (temporarily as it turned out) shrinking market. This was a precursor to the "Pac-Man" consolidation strategy promoted by President Clinton's then Deputy Secretary of Defense, William Perry, at a meeting with the defense titans, dubbed the "Last Supper." Perry's strategy led to industry-wide mergers in the early to mid 1990s. Significantly, when the defense budget began to grow rapidly after 1998, there was no undoing of the consolidation. Thus, today the defense industry is dominated by three giant all-purpose weapons manufacturers, two of which now have their headquarters in the Washington, D.C. area, and the third (Boeing) with a major government relations office in the D.C. area as well, to more closely supervise their most important corporate activity: the lobbying efforts that influence the money flow out of the Pentagon, Congress and White House.

Turning Clausewitz on His Head

It is easy to throw rocks at President Obama, but he did not create the defense mess, nor did his predecessor - though George W. Bush's reckless spending, coupled with incompetent management in Donald Rumsfeld's Pentagon and the domestic politics of the war-centric foreign and domestic policies that metastasized in the wake of Sept. 11, certainly worsened the crisis and accelerated the Pentagon's day of reckoning.

In fact, Mr. Obama inherited a Defense Department that was in the terminal stages of a meltdown first ignited as far back as the mid 1950s, when the unit costs of weapons started to grow substantially faster than the defense budgets. The deliberate explosion of military electronics spending - radar and other sensors, automation, communications, and then digitization - in the late 1960s greatly accelerated this cost growth and widened the mismatch further. That huge cost growth was (and still is) justified with a myopic argument, entirely tautological, that rising cost and technical complexity were a necessary consequence of our advantages in technology - and it was this technology that was the source of our strength.

The dogmatic belief that greater weapons system complexity and, even worse, greater organizational complexity enhances combat effectiveness is at the epicenter of the belief system sustaining the MICC. In truth, as later essays in this book will show, the out-of-control complexity of our weapon and command systems has shackled our forces in the field, making them rigid, predictable and highly vulnerable to faster thinking, more creative and more adaptive enemies using far simpler weapons and systems of command.[7] Our drive towards complexity makes a mockery of the hard-learned lessons of history going back thousands of years.

Most readers have heard of the KISS principle, distilled by World War II GI's out of their hard-won combat experience: Keep It Simple, Stupid. KISS and its antithesis, complexity, were hardly new concepts in the 1940s. They are, for example, at the heart of Clausewitz's 200 year-old theory of friction in combat - encapsulated in his famous statement that, "Everything in war is simple, but the simplest thing is difficult. The difficulties accumulate and end by producing a

[7] "Complexity (technical, organizational, operational, etc.) causes commanders and subordinates alike to be captured by their own internal dynamics or interactions; hence they cannot adapt to rapidly changing external (or even internal) circumstances." (Col. John R. Boyd, "Patterns of Conflict," slide 176) To understand the reasoning underlying this brilliant and original insight, download the entire "Patterns of Conflict" briefing at http://dnipogo.org/john-r-boyd/. Also, find an excellent and very readable biography of Boyd in Robert Coram's *Boyd: The Fighter Pilot Who Changed the Art of War* (Little, Brown and Company, 2002).

kind of friction that is inconceivable unless one has experienced war."[8] Clausewitz considered friction to be the pervasive atmosphere of war, or the fog of war; his musings on the proper conduct of war emphasized simplicity to reduce this friction.[9] The ideology of the American military - and its academies - purports to be grounded in Clausewitzian thinking. Yet, for at least the last 40 years, military service doctrine, headquarters briefings, and defense contractor brochures and propaganda have continuously asserted that increasing the complexity of our technology and organizations is the key to lifting the fog of war. The complexity dogma becomes ever more deeply ingrained, notwithstanding our painful combat lessons on the ineffectiveness of complex weapons and command systems in Vietnam, Kosovo, Iraq and Afghanistan.[10]

In the 1970s, 1980s and 1990s, using mounds of data and analysis, reformers clearly laid out the future consequences of the cost-budget mismatch in terms that were never rebutted empirically by the defenders of the status quo.[11] Despite that, the military reformers were unable to convince either the Pentagon leadership or members of Congress of the long-term dangers posed by the

---

[8] See: Carl von Clausewitz, *On War*, ed. Anatol Rapoport (Penguin Books, 1968). Chapter 7 in Book 1 discusses friction.

[9] According to Clausewitz : (1) each adversary possesses an independent will and therefore can act and react unpredictably; (2) uncertainty of information acts as an impediment to vigorous activity (i.e. friction); (3) a variety of psychological and moral forces can impede or stimulate vigorous activity; and (4) military genius can overcome friction, simplifying the myriad difficulties of war. These ideas are timeless but, as American strategist Col. John Boyd has shown, Clausewitz overemphasized the importance of reducing your own friction while greatly underestimating the importance of amplifying your adversary's friction. See slides 40, 41 and 42 of Boyd's "Patterns of Conflict," downloadable at http://www.dnipogo.org/boyd/pdf/poc.pdf, or http://www.dnipogo.org/boyd/patterns_ppt.pdf.

[10] While the lessons of Vietnam, Iraq and Afghanistan are clear, some readers may question the inclusion of Kosovo. Kosovo is a case study in the failure of high complexity weapons and organizational arrangements. U.S. military planners predicted a "precision" bombing campaign would force the Serbs to capitulate in only two to three days, but the air campaign grinded on for 79 days. Yet when it was over, NATO intelligence determined only tiny quantities of Serbian tanks, armored personnel carriers, self-propelled artillery, and trucks were destroyed. Serbian troops marched out of Kosovo in good order, their fighting spirit intact, displaying clean equipment, crisp uniforms, and in larger numbers than planners said were in Kosovo to begin with. Moreover, the terms of Serb "surrender," which the undefeated Serb military regarded as a sell out by Serbian President Milosevic, were the same as those the Serbs agreed to at the Rambouillet Conference, before U.S. negotiators and Secretary of State Madeline Albright inserted a poison pill to queer the deal, so we could have what the politically troubled Clinton administration thought would be a neat, short war.

[11] Readers interested in examples of the numbers and logic behind this statement are referred to (1) *Defense Facts of Life* (Westview, 1985), (2) "Defense Time Bomb," (http://pogoarchives.org/labyrinth/01/07.pdf), and (3) "Defense Death Spiral," (http://pogoarchives.org/labyrinth/01/05.pdf).

increasing complexity/ineffectiveness of our hardware and organizations, the shrinking, aging force consequences of the weapons cost explosion, the combat dangers posed by the rigid, techno-dependent mindset, or the corrosive influence of the warped financial incentives that fueled this death spiral.

In response, the factions of the MICC united in persuading a succession of presidents to waste 30 years pursuing the fantasy that we could buy our way out of the military-economic death spiral with ever larger defense budgets funding fewer numbers of ever more complex and costly weapons. The circularity of the underlying argument for complexity was perfectly expressed in 1980 by Defense Secretary Harold Brown, a leading proponent of high-tech spending and one of the chief architects of the shrinking, aging force: "Given our disadvantage in numbers, our technology will save us."[12]

A telling vignette of the buy-our-way-out fantasy is the Ronald Reagan spending spree beginning in 1981: his budget increases unleashed a round of cost growth wherein weapons costs grew at a far faster rate than ever before, thereby widening the gap between accelerating unit costs and the much slower growth of the overall budget. Those Reagan budget increases led directly to a 1990 combat force structure that, overall, was smaller and older than in 1981. Similarly, the ongoing Clinton-Bush-Obama spending spree that began in 1999 merely set the stage for a today's much larger crisis.[13]

A Case in Point: The 2010 QDR

We decided to produce this anthology in early 2010, when it became clear that President Obama's defense team was not up to even acknowledging, let alone fixing the core problems discussed above. This became obvious when the Pentagon released the results of the Quadrennial Defense Review early in 2010, one year into Mr. Obama's presidency.

When a new president assumes office, as Mr. Obama did in January 2009, he inherits the long- range defense budget plan that was produced over the preceding 18 months by his predecessor's Pentagon. Given the reality of a Congress committed to ongoing spending programs, there is little any president can quickly do to change his predecessor's budget in a way that reflects his own policy intentions, unless he just wants to indiscriminately throw money at the

[12] As quoted in *Newsweek* in 1980; see David Dickson, *New Politics of Science* (University of Chicago Press, 1988), p. 125.
[13] Readers can confirm this by referring to the cost, force structure and age data in the last two references of footnote 11.

Pentagon, as Ronald Reagan did in 1981.[14] In effect, the new president is a prisoner of the Pentagon's fatally flawed bureaucratic planning process known as the Planning, Programming and Budgeting System (PPBS)[15] and all the MICC budget games it contains.[16]

But there is more. Mr. Obama also inherited a congressionally-mandated requirement to produce a long range strategy document during his first year in office. This document is known as the Quadrennial Defense Review or QDR, and it is required by law, every four years, at the end of the first year of a newly elected president's term.[17] The QDR is supposed to shape the activities of the PPBS, but they both go on simultaneously, and by necessity, pretty much independently. Nevertheless, the 2010 QDR was Mr. Obama's first real chance to imprint his policy intentions on the MICC.

Obama's Pentagon let him down by producing yet another sham of a QDR.[18] To make a long story short, consider just one important example. Judge for yourself if it suffices to make the point.

First, a little background: the Pentagon has been producing FYDPs since 1962. But they have been repeatedly criticized, quite rightly, for producing defense

---

[14] In 1981, the Reagan administration was so intent on throwing money at the Pentagon they chose to rush through an amendment to President Carter's 1981 budget. Without any kind of systematic review—and not having the time to type up a new budget—Reagan's political appointees directed the Pentagon to just write in pen-and-ink changes adding billions of dollars to hundreds of line items. Much of this largesse was immediately converted into cost growth in existing programs.

[15] The product of the PPBS is the Future Years Defense Plan or FYDP. This document is produced by staff work involving millions of man hours over a period of 18 months; it lays out the Pentagon's future spending intentions for the next five years for thousands of individual line items. The first year of the FYDP is the budget that is sent to Congress each February. So, with only two months to make changes, and a staff not fully in place, the most a new president can do is make a few marginal changes to his predecessor's document.

[16] A description of the MICC's gaming pathologies can be found in my 1991 pamphlet "Defense Power Games," (http://pogoarchives.org/labyrinth/01/09.pdf) and in my June 4, 2002 statement to the Subcommittee on National Security, Veterans Affairs and International Relations, Committee on Government Reform, United States House of Representatives (http://pogoarchives.org/labyrinth/01/02.pdf).

[17] In the 1994 National Defense Authorization Act, Congress mandated the Commission on Roles and Missions (CORM). Among the usual plethora of "feel-good" recommendations was the idea that DOD should undertake a major quadrennial strategy review. Reacting to this, Congress directed the 1997 Quadrennial Defense Review as a method to conduct a "fundamental and comprehensive examination of America's defense needs."

[18] See, for example, my critique of the first QDR in 1997, which can be downloaded from http://pogoarchives.org/labyrinth/01/06.pdf.

budgets disconnected from the national military strategy. Because the dollar allocations made in the budget define the government's real policy, the critique was logically equivalent to saying there was no functioning national strategy, and budget decision-making was actually driving strategy (which was and still is the case). The QDR legislation was the most recent attempt to deal with this long-standing criticism by requiring the Pentagon to lay out a framework for matching its military strategy and policy ambitions to its budgetary, people and technology constraints.

The 2010 QDR, together with the new FY 2011 budget (and accompanying FYDP), therefore, are supposed to permit an analysis of the strengths and weaknesses implicit in the administration's proposed match-up between resources and strategy. This information would then become the grist for a rational national debate by linking strategic considerations to the inevitable compromises made in the sausage-making factory that is Congress. Moreover, as this was the first defense budget President Obama controlled from beginning to end, and because it represented $700 billion-plus that Mr. Obama had temporarily put off limits in the extant debate over spending, it was crucially important for the Pentagon to get the QDR and the accompanying FYDP right in a logically consistent and transparent manner.

The Pentagon flunked the test.

For the past 20 years or so, the mainstream press, the Government Accountability Office (GAO) and the Pentagon's own Inspector General have inundated the American public with well-supported horror stories about the Pentagon's aging and shrinking force structure, the Pentagon's unauditable budget shambles, the apparently deliberate inability of the Pentagon's accounting system to track actual expenditures, the weapons cost growth that outstrips the budget growth and, more recently, the wear-out of the force structure caused by our never-ending wars, and the alarming increase in Post-Traumatic Stress Disorder (PTSD) casualties (and suicides) caused by the excessive troop rotations mandated by shrinking force structures.

With this in mind, readers should now download the QDR and the FY 2011 Budget Overview from these links:
QDR -
http://www.defense.gov/qdr/QDR%20as%20of%2029JAN10%201600.pdf
FY 2011 Budget Overview –
http://comptroller.defense.gov/defbudget/fy2011/FY2011_Budget_Request_Overview_Book.pdf

These reports are searchable PDF files. I urge readers to do word and phrase searches on terms like "age," "weapons aging," "shrinking forces," "weapons cost growth," "wear-out," "excessive troop rotation," "sustainability of deployments," "accounting," "audit," "tracking expenditures," or anything else one can think of that might relate to the widely known and overwhelmingly important people, money and hardware problems described above. Determine for yourself that not one of these vital national security issues are acknowledged, addressed or analyzed in either the QDR or the defense budget.

A search in the "Budget Overview" document for the word "audit," for example, will take you to page 7-34, among others, where you will find that DOD set a goal of reaching 100 percent readiness to audit its assets and liabilities by the year 2017, but the last column shows that the indicator of progress made toward that goal in FY 2010 was deleted at the request of the Comptroller, who happens to be the chief financial officer of DOD! Furthermore, the "auditability" asserted for 2017 would not include any audits of specific weapons programs!

Alternatively, you could search for any mention of these central problems by reading the entire text - but be advised, it makes for grim reading. Either way, a reader that approaches this task objectively will end up with the same conclusion: these documents fail to touch on any of the pressing strategy and resource problems described above, much less present plans for correcting any of them.

Defenders of the MICC status quo might say we must go forward with these ridiculous plans that do nothing but whitewash business-as-usual because we are at war and need the resources "for the troops." But that argument merely proves our point about the MICC needing continual war to keep its political economy afloat.

Nor is it true that the ongoing wars force us to accept the budget as is: President Obama could freeze the non-war defense budget at this year's level, just like he is doing for the rest of discretionary spending by the government. He could tell the Pentagon to go back to the drawing board and produce some plans that address the all-too-real fiscal problems we face. He could declare the bookkeeping shambles a task of the highest national security urgency - which it is - and order the Pentagon to clean it up with a massive crash program, leaving the budget freeze in place until full and complete auditability is achieved, or better yet conduct the audits themselves.

The omission of critical thinking, the failure to engage DOD's most crucial problems in the 2010 QDR is no accident - it represents a defense of business-as-usual. And business-as-usual brings us full circle back to Colonel Boyd's quote at the beginning of this introduction: the Military - Industrial -

Congressional Complex *"...does have a strategy; it is: don't interrupt the money flow; add to it."*[19] The QDR is the handmaiden of that strategy.

Pressing On

The rest of this handbook is concerned with providing the reader the tools for assessing national strategies that serve the country's interest rather than the MICC's - and for assessing productive changes in the money flow, changes that contribute to improved training and better combat leadership for our people in uniform, more effective weapons that cost less, cures for the shrinking and aging forces, full accountability throughout DOD, and sustaining troops in the field without the excessive rotations that incur widespread psychological—and physical—casualties in wars now driven more by the need to keep the MICC afloat financially than by considerations of the national interest.

Our aim in all this can be found in James Madison's call for an informed citizenry: "A popular government without popular information, or the means of acquiring it, is but a prologue to a farce or a tragedy, or perhaps both. Knowledge will forever govern ignorance, and a people who mean to be their own governors must arm themselves with the power which knowledge gives."[20] In that spirit, we hope to provide enough background and orientation to enable our readers to determine for themselves what has gone wrong and to assess what might be needed to end America's defense meltdown.

---

[19] For more in summary fashion about Colonel Boyd, see my summary of his life's work at http://pogoarchives.org/labyrinth/01/01.pdf.

[20] James Madison, from a letter to W.T. Barry, August 4, 1822, at http://press-pubs.uchicago.edu/founders/documents/v1ch18s35.html.

Essay 2

"Penetrating the Pentagon"

by George C. Wilson

The marching orders Executive Editor Benjamin C. Bradlee of *The Washington Post* gave me back in 1966 when he hired me to cover national defense for his newspaper are even more important for editors, reporters, congressional staffers, lawmakers, the secretary of defense and his deputies, and even the president of the United States to follow today than they were then:

> "The Pentagon spends all our f------ money but we never get a story out of there. Go break some loose. Find out where all that money is going."

Following the money pays dividends not only to editors and reporters, which was Bradlee's interest, but to officials in Congress, the Pentagon, the White House, the defense industry and think tanks scattered around the country. Counting the money going for homeland defense and nuclear weapons, the total amount the president spends each year to protect Americans at home and pursue foreign policy objectives abroad with military force, is more than $1 trillion. Yet we spend much more today than during the Cold War with the Soviet Union and its Warsaw Pact, which compelled us to buy and deploy forces roughly twice the size of today's. Two wars the United States is fighting in Iraq and Afghanistan are against foes with no standing army, no navy worth worrying about and no air forces.

Where is all that taxpayer money going and why? Those are questions that persons in the news media, Congress, the Pentagon and White House should be asking every day. Sadly, Congress has all but forfeited its powers written in Article 1, Section 8 of the Constitution to "provide for the common defense" and "to declare war." Not since Congress declared war in 1941 in response to the Japanese attack on Pearl Harbor have the lawmakers, hired hands of voters living in their states and Congressional districts after all, exercised those Constitutional powers. The legislative branch since 1941 has allowed the executive branch, in the person of both Democratic and Republican presidents, to send young American men and women into battles abroad where many tens of thousands have been killed and wounded. Congress passed the War Powers Act in 1973 in hopes of getting back some of the powers it foolishly gave away to the president. But this has not happened. In my view, the lawmakers are guilty of malfeasance or nonfeasance, but few in the government or media are demanding accountability.

Few Americans realize that the United States has military forces in 150 different countries around the world. The Pentagon puts out a press release every year disclosing this far-flung presence and the number of service people in each country. But only a few Pentagon watchers notice this global reach of the U.S. Army, Navy, Air Force and Marine Corps. Even fewer question, challenge or think about the implications of this extensive American military presence abroad. And I doubt if the Pentagon's press release on foreign deployments includes American commandos in uniform and the CIA's hired guns who are all around the world assassinating suspected terrorists and hostile tribal leaders. It is only a matter of time in my view before these raids on the ground and from the air provoke retaliatory attacks on the American homeland. If this indeed happens, Americans will find themselves with less freedom of movement and will probably have to carry identification papers with them at all times.

One lawmaker who has indeed thought and worried about our extended American military presence abroad is Chairman Bob Filner, D – Calif., of the House Committee on Veterans' Affairs in the 111th Congress. This is what he is worried about, as recorded in an interview I had with him in his Washington office in 2010:

"I was trained in ancient history, and I taught a lot of ancient history. The Athenian Empire always struck me as a parallel to our situation today. They started off as a democracy and because they expanded and took over other countries, what happened at home killed them. They ended up losing everything."

Ever since Sept. 11, the United States has been in what some defense specialists term "The Long War" against terrorists and terrorism. Retired Army Col. Andrew J. Bacevich, a professor at Boston University and author of several books decrying our overextended military, is representative of those who believe the United States is shooting itself in the foot. On the other hand, John O. Brennan, President Obama's counterterrorism advisor, is among those who contend the United States as a matter of self-defense must go after terrorists wherever they show themselves. The poles of their argument:

> Bacevich. "For the United States after 9/11," wrote the soldier-scholar in *The Limits of Power,*[1] "war became a seemingly permanent condition. By and large Americans were slow to grasp the implications

---

[1] Andrew J. Bacevich, *The Limits of Power: The End of American Exceptionalism* (Metropolitan Books, 2008). See also Bacevich's new *Washington Rules: America's Path to Permanent War.*

> of a global war with no exits and no deadlines. The United States is ill prepared to wage a global war of no exits and no deadlines. The sole superpower lacks the resources—economic, political and military—to support a large scale, protracted conflict without, at the very least, inflicting severe economic damage on itself....Seven years into its confrontation with radical Islam, the United States finds itself with too much war for too few warriors—and with no prospect of producing the additional soldiers needed to close the gap. In effect, Americans now confront a looming military crisis to go along with the economic and political crises that they have labored so earnestly to ignore."
>
> Brennan. In a little noted speech cleared by the White House and presented at the Center for Strategic and International Studies on May 26, 2010, said this: "The United States of America is at war. We are at war against al Qaeda and its terrorist affiliates...We will not merely respond after the fact—after an attack has been attempted. Instead, the United States will disrupt, dismantle and ensure a lasting defeat on al Qaeda and violent extremist affiliates. We will deny al Qaeda and its affiliates safe haven...To deny al Qaeda and its affiliates safe haven, we will take the fight to al Qaeda and its extremist affiliates wherever they plot and train...We often need to use a scalpel not a hammer. When we know of terrorists who are plotting attacks against us, we have a responsibility to take action to defend ourselves, and we will do so."[2] (The scalpel reference, according to military officials, includes assassinations from the air by armed unmanned aircraft, as well as by commando teams on the ground striking suspected terrorists at night.)

Bacevich sees the American part of the global war on terror as a bridge too far while Brennan is determined to cross it, no matter what the cost nor for how many years this "Long War" must be fought. These conflicting views will surely split the American government and body politic in the years ahead and perhaps the U.S. military as well.

The challenge for the media, members of Congress, their staffs and committees, is to see these cracks as they develop and describe them and their implications to the American public. There is a huge elephant in America's living room. Feeling and explaining its various parts accurately and clearly will be a challenge.

What follows are suggestions from a veteran defense reporter on how to learn about this new elephant in the American living room and explain what damage it

---

[2] Find these comments at the White House Web site at http://www.whitehouse.gov/the-press-office/remarks-assistant-president-homeland-security-and-counterterrorism-john-brennan-csi.

could do to our country. Even if you support the "permanent war" as now being conducted, examining it and its likely consequences will be enlightening.

Understanding Congress

Staffers. There are more than 40 congressional committees and subcommittees that delve into some facet of Pentagon business. Each of those committees and subcommittees has a staff whose members should—but don't always—know more than the chairman and ranking member. Nevertheless, Washington is a top-down bureaucracy, and it behooves the would-be investigator of the defense establishment to tell the chairman and ranking member that you would like to talk his or her staff on background (meaning not putting the staffers' names in print) to learn the nitty-gritty of an issue. The typical chairman and ranking member will breathe a sigh of relief for not being questioned themselves, and pass the word that his or her staff should talk to the investigating individual.

Once the chairman or ranking member has blessed the inquiry, committee and subcommittee staffers will usually feel free to share their insights about the issue at hand, especially if you spend face time with them. Committees and subcommittees have separate staffs to serve the majority and minority members. So be sure to talk to both sides.

One of many advantages of doing this on Capitol Hill is that politicians by nature are vocal while bureaucrats in the Pentagon are by nature close-mouthed. Internal Pentagon studies you hear about but are not released can often be obtained through the good offices of a representative or senator.

Congressional hearing transcripts. For the patient but hopefully speedy reader there are nuggets of information to be found here. Most newspaper and TV reporters who attend hearings or breeze through their transcripts are looking for exchanges that advance the military story of the moment, not provide insight on strategy nor obscurely described flaws in expensive weapons. Former Secretary of State Condoleezza Rice, for example, at a Senate Foreign Relations Committee hearing spelled out the Bush administration strategy for Iraq – "clear, hold and build." The mainstream press in its hurry to find a fight or forecast allowed this attributed quote to lie dormant for days before putting it in print.

Congressional Research Service. Bright people write CRS reports after weeks and often months of research. Their work is screened by their bureaucratic bosses who do not want to offend anybody. The authors of these reports have to stand by while their bosses pull out the sharpest teeth. Still, the work is substantive and provides insights for serious students of the military-industrial-political-intelligence complex. Many of the CRS analysts have been around for years, a relationship of trust with them can earn you what did not show up in the final version of the report. Perhaps a trusting researcher will alert you to, or even give you if the bond is strong enough, a penetrating but enlightening, unclassified report which a bureaucrat hid from public view.

Government Accountability Office. GAO is a creature of Congress and does studies for it which constitute the most accessible and quotable bean-counting reports on Pentagon spending available to lawmakers, staffers, pressure groups, Pentagon bureaucrats, generals, admirals and interested citizens. GAO procedures require its auditors to obtain comment on their findings from the Pentagon before releasing them to the public, often defeating GAO's attempts to be timely. As with CRS reports, the findings are often ground down to mere hints of scandalous misuse of taxpayer dollars. Sometimes a friendly GAO insider will slip you the draft report with the teeth intact.

To get the full story of a horrendous cost overrun on a weapon, you will have to dig up the various pieces from contract award to flawed final product to arrest the attention of the lawmaker, the Pentagon or the public. Good contacts at CRS and GAO will help you do that.

Ambushing Senators and Representatives. If you need a quote in a hurry for a story or report, the best way is to set up an ambush of the lawmaker rather than go through the bevy of young blonde press secretaries in today's congressional offices who can be decidedly unhelpful. Productive ambush positions for senators are the escalators just above the Senate subway under the Capitol building. House members can be ambushed near their subway, too, but many walk outside and find other ways to elude an ambush. If you have press credentials, you can fill out a card asking the senator or representative to leave his or her desk and talk to you in the lobby outside the Senate or House chamber. I confess to missing the old days when a senator or representative whom you called off the floor trusted a reporter enough to speak in his native tongue, knowing you would leave out his swear words in writing up his remarks.

One of the most direct senators was Sen. Barry Goldwater, R – Ariz. If he trusted you, he would tell you what he really thought about an issue or a person, sprinkling his responses with four letter words. Former Deputy Defense Secretary John Hamre told me that when he was a staffer on the Senate Armed Services Committee he and others there drafted a letter for Goldwater only to sense he was unhappy with it. "It doesn't sound like me," Goldwater complained. "Throw some s---s and f---s in there and it'll be all right."

I called Rep. L. Mendel Rivers, D – S.C., off the House floor one day to ask the senior member of the House Armed Services Committee how he had persuaded liberal Representatives from New York to vote for money for weapons they had publicly opposed. "What you're really asking me, Jawrge, is how I got the wops and Jews to vote for my bill. I told them if they didn't vote for it, I'd go into their home districts and campaign for 'em."

Luncheon or evening meetings. These are where lawmakers give speeches and answer questions. Members of the mainstream press seldom attend these events, but they can be fruitful, especially the unrehearsed and unvetted question and answer periods. Among the places where such appearances are listed have been *CongressDaily* and *Congressional Quarterly*. You can usually get face time with lawmakers who during the day are protected by horse-holders.

The Essential Field

By far the best place to learn about the American military is not inside the Pentagon but out in the field where soldiers, sailors, airmen and marines train or fight. A credentialed news reporter or congressional staffer can get that learning experience by asking the heads of Army, Navy, Air Force and Marine Corps information offices in the Pentagon to let him or her see our military forces in action, including in combat if the newspaper, magazine or TV station will sponsor the news person and pay the bills. Congressional staffers can ask the military liaison officers on Capitol Hill to arrange similar forays.

One of the many dividends to embedding in an active duty military outfit is the mutual defrosting that will occur, especially if the embedded person stays overnight and is otherwise living with the troops. Officers will see for themselves that the news person or congressional staffer is trying to learn their business, not trap them into saying something that will get them fired as happened to Army Gen. Stanley McChrystal after hosting a *Rolling Stone* reporter. I do not know what the ground rules were, but generals and their deputies often criticize politicians if they believe the senator, representative, congressional staffer or embedded reporter will not attribute their critical remarks to them by name.

Also, winning a military officer's trust and respect in the field will give the reporter or staffer a knowledgeable person to telephone, e-mail or visit months later in Washington when you are trying to understand a problem or unravel a scandal or figure out why cost overruns on weapons are so high. Such post-visit contacts are usually more enlightening if the military person's views are set forth without using his or her name. Anyone who tries to penetrate the military-industrial-political-intelligence complex needs a board of advisors who have been there, done that.

I found in the almost 50 years that I covered the American military that much of what they did and thought, and why, was fascinating. I wrote six books on the human side of the military while covering it as a reporter, mostly for *The*

Washington Post. In that time I humped around with combat troops in Vietnam, Bosnia, Panama, the Middle East and Iraq; sailed in warships and submarines, deployed seven and a half months on the aircraft carrier *John F. Kennedy* and flew as a side- or back-seater in every plane on her deck. The education about the military one gets in the field can be monumental and invaluable.

Inside the Pentagon

Secretaries of defense who open new vistas during an interview are rare. Defense Secretary William Perry was one who did; Harold Brown could do so if he felt like it. I think Defense Secretary Robert Gates talks a better game than he plays. He did not cancel the $350 million per copy F-22 fighter program, for example, just the last few on order. He has written and spoken about the excesses in the Pentagon he is supposed to rule. But he has said repeatedly that what he would like to do and what Congress would let him do are two very different things. As a result, nothing can happen. When all is said and done, a defense secretary works for the president and cannot go beyond his wishes on major issues like canceling a weapons program.

The secretary's deputies vary in their willingness to open the kimono to you on strategy or even what they are working on at any moment. Heretics inside the Pentagon bureaucracy can be found and interviewed much more productively but usually do not like to be quoted by name.

E-mail can be a useful weapon for a reporter or congressional staffer. Getting into electronic conversations with knowledgeable people inside the Pentagon can be enlightening, especially if they understand their names will not show up in your article or report and they know they can trust you on that. A contact list full of e-mail addresses is better than a little black book full of telephone numbers nowadays. Queries to generals, once you have their e-mail addresses, can be especially productive.

Occasionally, someone really senior, like a member of the Joint Chiefs of Staff, will figure it is in his interest to establish a dialogue with a member of the press or congressional staffer. When I was covering the Pentagon for *The Washington Post,* I had breakfast monthly with a member of the Joint Chiefs of Staff and could usually get in to see its chairman. It was helpful seeing issues facing the armed services and the country through his end of the telescope.

One chairman of the Joint Chiefs unburdened himself to me about the folly of spending billions on missile defense when a lone terrorist on a cruise boat circling Manhattan could kill hundreds and perhaps thousands of New Yorkers at lunch time just by lobbing a mortar or two with poison gas or bacteria-germs in them into the crowded streets. "The terrorist could just board the boat with the mortar tube under his coat, set up the mortar on the rear deck in seconds, launch the mortars and then throw the tube overboard," the four-star general told me. You can get more insight from such relationships.

Information officers can be useful but vary greatly in willingness to be enlightening, or even really helpful. The top information officer is usually more informative than his deputies, so it can be productive to build bridges to him or her. Navy Capt. Jay Coupe when he was top information officer for Adm. William J. Crowe Jr. in the mid-1980s could be informative, funny and savage on policy and people by turns. But he had the friendship and trust of Crowe to a degree spokesmen before and after could not duplicate, only envy. Most spokesmen for the chairman try to glorify him rather than be informative on the issues before the Joint Chiefs.

Think Tanks and Scientists

Reporters and staffers who keep in touch at least by telephone or e-mail with both the hawks and doves who roost in Washington think tanks, foundations and universities are well served. During the reign of Robert S. McNamara as secretary of defense, I broke the story on page 1 of *The Washington Post* on Jan. 29, 1967 that the United States had perfected the technique of packing several nuclear bombs into the nose of one missile and sending each of them to cities hundreds of miles apart.

I did not meet a pre-Deepthroat official on a dump at midnight to get that story. I listened instead to arms control scientists at one of their annual meetings in New York City where they said an adversary could overwhelm any missile defense the United States constructed with real and dummy warheads. The smarter and cheaper alternative to a missile defense was to build a missile offense which could fling so many warheads at the enemy that the enemy could not stop every one of them. Such an offense would inspire both American and Soviet leaders to rely on what became known as Mutual Assured Destruction, or "MAD." Both superpowers ended up on relying on this "I won't if you won't" form of deterrence rather than wipe each other off the face of the earth with nuclear bombs during the coldest part of the Cold War.

As I listened to the exchanges between scientists, I sensed the urgency about the need to learn how to pack several nuclear bombs into the nose of one missile rather than keep digging holes for single warhead missiles in the American West. I learned at the same unclassified scientific meeting that the desired technology was called MIRV for multiple-independently-targetable re-entry vehicle.

Back at *The Washington Post* after the meeting in New York, I called the arms controllers and scientists I knew and got a nugget here and there which advanced my knowledge of what was then to become the great MIRV race between the United States and Soviet Union.

I wrote a news story about everything I had learned about MIRV and showed it to *The Washington Post* editors J. Russell Wiggins and Bradlee. I knew printing such a sensitive story about secret technology would trigger howls of protest from McNamara and other Pentagon leaders. Wiggins and Bradlee debated in my presence whether to print the MIRV story or not. "It's your country, too, Ben," Wiggins lectured Bradlee at one point. Wiggins insisted I brief Pentagon leaders on my MIRV findings to obtain their reaction and arguments against printing it. Bradlee reluctantly went along. His mantra was "the name of the game is impact." And the MIRV story I had written would indeed have impact.

My first stop at the Pentagon with the story in hand was Arthur Sylvester, head of the Department of Defense information directorate in 1967. Sylvester read my MIRV story and immediately got on the telephone line which rang directly on McNamara's desk. "I've got George Wilson in my office and I just read the story he gave me," Sylvester told McNamara. "It's got a lot of MIRV in it." Sylvester told me that McNamara had replied, "You know as well as I that any story discussing MIRV would be harmful. Have George talk to Harold Brown" who at the time was the Pentagon director of defense research and engineering.

Brown told me that "we would rather have you print no story at all." But he added that if *The Post* insisted on printing some kind of MIRV story, it would be much less harmful to the national interest if our paper left out the fact that the United States had perfected the technique of having a single missile independently drop one nuclear warhead on Soviet cities hundreds of miles apart as its "bus" flew along. The Soviets already knew, Brown said, that the United States could pack several warheads into one missile and drop them shotgun pattern on a single target area, not a series of them hundreds of miles apart.

Back at *The Post* I passed on to my editors the objections of McNamara and Brown. I felt then and now that the real breakthrough MIRV had achieved was obviating the need to keep on digging more and more missile silos. Shortly after John F. Kennedy was elected president in 1960, McNamara himself called reporters into the defense secretary's office complex and said on background that the "missile gap" Kennedy had accused the Eisenhower administration of opening up did not exist at all. It was a devastating story to Kennedy at the time. I argued before Wiggins and Bradlee that the MIRV breakthrough was all the more reason to stop digging missile silos because the missile gap alleged by Kennedy was a fraud. We had to describe high up in the story that one U. S. missile in a silo or inside a submarine could destroy a number of Soviet cities hundreds of miles apart because of MIRV. I leaned the Pentagon's way by leaving out of the final draft of my MIRV story several technical details about MIRV that might help the Russians better understand the system. My lead on the MIRV story was: "The United States knows how to use a single missile to destroy several different cities or military bases spaced hundreds of miles apart."

I thought then and now that *The Post* had acted responsibly. It had heard out the objections of Pentagon leaders to the MIRV story before printing it; considered their objections along with the need to enlighten its readers on the significance of the technological breakthrough and left out technical details that might help the Russians develop their MIRV.

But McNamara soon displayed his pettiness by pretending his own MIRV story was brand new in an interview with *Life Magazine* printed on Sept. 29, 1967 and by ignoring the pains *The Post* editors took to act responsibly in publishing a story he himself had said was sensitive. This is what McNamara told the *Life* interviewer months after my MIRV story had run on page one of *The Post*:

> "We're capitalizing on a major new technological advance. We can now equip our boosters with many warheads, each of which can be aimed at a separate target. We call this MIRV – Multiple Independent Re-entry Vehicles.
>
> "Q. Does the public know about MIRV?
>
> "A. There have been allusions to it in the press, but it has not been described publicly…"

The so-called "allusions" McNamara referred to was a column one story in *The Post* which ran in about 1 million Sunday newspapers printed eight months before his *Life* magazine interview was published.

I found during my years at *Aviation Week & Space Technology* magazine and later at *The Post* that magazines and newspapers who act responsibly on sensitive stories usually end up getting bitten by the very same government officials whom they sounded out before leaping into print. But despite being burned by McNamara, I still think responsible newspapers, television stations and internet outlets have an obligation before printing sensitive stories to ask relevant government officials what is the worst that would happen if the story at issue were published or televised.

I discovered during my five years at *Aviation Week* that the defense industry is an under used source of informative and often exclusive stories. One reason many reporters do not develop sources within the defense industry is that they are self-conscious about their shallow understanding of how complicated weapons work. Almost every news person has covered politics at the local, county, state or federal level, but few similarly go inside defense plants where engineers can explain, often with models, what they are working on and why.

The trick is to simply tell information officers at these companies that you want to better understand the weapons they have sold and hope to sell to the Pentagon and military services. Most information officers see it in their company's interest to arrange briefings for the reporter with knowledgeable engineers and scientists. *Aviation Week* had refined the process to the point that editors and reporters received three days of elaborately arranged briefings at aerospace firms all over the United States. The vice president in charge of the Washington office of aerospace firms and his team, usually including retired admirals and generals, can be a source of all kinds of enlightening information on what is going on behind closed doors at the Pentagon and in Congress once the reporter or congressional staffer establishes rapport with them.

As I said earlier, there is a giant elephant in America's living room – namely the military-industrial-political-intelligence complex. I think it is out of control. Congress needs to grab back its Constitutional powers to provide the common defense and to declare war. To wrest those powers from the president, the elephant first has to be felt all over by congressional committees, the media and groups worried about where our militarism is going and why. The dangers of this elephant to America's future need to be described clearly, with excessive spending on weapons and provocative military deployments just two of many dangers. If this elephant is not brought under control soon, it will trample all over our beloved and envied democracy.

Essay 3

"Learning About Defense"

by Bruce I. Gudmundsson

The layman who wants to make sense of some aspect of defense will find a number of obstacles in his way. In particular, he will encounter problems of language, culture and pecuniary interest. These obstacles are daunting, but not insurmountable. Indeed, with a little bit of effort and a reasonable amount of background information, a young professional, even an ordinary citizen, can gain a sufficient understanding of any given corner of the defense establishment to determine whether a particular decision, idea or project is in the public interest.

The purpose of this essay is to give the layman, whether journalist, congressional staffer or interested citizen, a way to make sense of the vast defense establishment that has rooted itself in the American body politic in the last century or so. It does this by using the simile of the island of New Guinea. Like New Guinea, the defense establishment is both obvious and mysterious, a place with both a thoroughly mapped coastline and an unexplored interior. Similarly, both the defense establishment and New Guinea are home to a large variety of tribes, each of which has its own language and culture, and each of which interacts with other tribes in a variety of ways.[1]

New Guinea is home to more than a thousand distinct dialects, the vast majority of which are peculiar to the mountainous interior of the island. As is often the case in places of such linguistic diversity, most of the inhabitants of the New Guinea highlands are bilingual. When dealing with outsiders, a New Guinea highlander uses a *lingua franca*, a common language of trade and travel. Within his own tribe, however, he speaks a tongue that is often completely unintelligible to people who live but a few miles away.

---

[1]The idea that the defense establishment is made of several distinct cultures is elegantly laid out in Carl Builder, *Masks of War: American Military Styles in Strategy and Analysis* (Johns Hopkins University Press, 1989). However, while Builder deals with the organizational cultures of each of the armed services – the Army, Navy, Marine Corps and Air Force – this essay lays out the proposition that smaller entities within the uniformed services, as well as private corporations and civilian agencies, also have peculiar cultures of their own.

In the defense establishment, the counterparts of the tribal languages of New Guinea are the jargon-filled, acronym-intensive dialects spoken within particular branches, commands and occupational fields. Some features of these "tribal" dialects are the result of deliberate action on the part of authorities. The United States Marine Corps, for example, devotes hundreds of thousands of man-hours each year to ensure that new recruits use new words (such as "bulkheads," "hatches" and "ladderwells") for things that already have perfectly good names in standard American English ("walls," "doors" and "staircases"). For the most part, however, the dialects of the defense establishment have evolved as all languages do. That is, new words are created, imported or repurposed to fill new needs, while words that fail to find regular employment are quickly forgotten.

Most of the new words coined for use by various communities within the defense establishment are acronyms. The practice of assembling new words out of the most conspicuous fragments of existing words is nearly as old as the alphabets that make it possible. Thus, acronyms have long played a part in the language of a wide variety of human communities. Few other realms, however, can compete with the American defense establishment when it comes to the number, variety and pervasiveness of such synthetic words. Indeed, we have reached a point where there are communities within the defense establishment that use acronyms made up of other acronyms. Thus, the acronym for the School of Marine Air-Ground Task Force Logistics is not, as one might imagine, "SMAGTFL." Rather, because "Marine Air-Ground Task Force" is an acronym in its own right (MAGTF), the school is universally known as "SOML." (The shorter acronym is pronounced "saw-muhl," and thus not nearly as much fun to say as "smag-tah-full.")

One reason for the popularity of acronyms within the defense establishment is the ease with which they can be coined. Unfortunately, things that are easy to make are also easy to discard, and so the shelf life of most acronyms tends to be rather short. This, in turn, increases the difficulties that face a person who is trying to learn the dialect of a particular community. In particular, the rapid turnover of acronyms reduces the utility of the lists of acronyms that are compiled from time to time. It also means that there are many people within defense communities who cannot spell, let alone identify the component words of, the acronyms that they use on a daily basis.[2]

To further complicate matters, different communities sometimes use the same acronym to mean entirely different things. Some tribes use "IW" to mean "information warfare." Others use the same pair of initials to mean "irregular

---

[2]Some of the larger communities within the defense establishment will periodically publish glossaries of acronyms and other terms of art. Many of these, including the mother of them all (*Joint Publication 1-02, Department of Defense Dictionary of Military and Associated Terms*), are available online.

warfare." Moreover, even when the component words of an acronym are the same, the meaning can be different. Thus, for example, an Air Force "FAC" ("forward air controller") is a person in an airplane, while a Marine Corps "FAC" (also "forward air controller") is a man on the ground.[3]

The obstacles to communication created by acronyms and other terms of art are as daunting to people within the defense establishment as they are to those outside of it. Because of this, communication between communities is usually conducted in the same *lingua franca* that is used for communications with the outside world. Indeed, when a person within the defense establishment uses plain English, it is usually a sign that he is attempting to communicate with outsiders of one sort or the other. (The other explanation is that he is one of those brave souls who have taken it upon themselves to resist the irresistible tide of linguistic diversity.)

As is so often the case with the languages of trade and travel, the *lingua franca* of the defense establishment is also the language of prestige. Thus, the plainer the English spoken by a person within that world of defense, the greater the chances that he is near the top of the local hierarchy. Because of this, the many schools that serve to groom people for senior leadership within the defense establishment place a great deal of emphasis on the ability of their graduates to read, speak and write standard American English. However, as this effort rarely results in complete fluency, many of the documents produced for intertribal and external consumption are the work of professional scribes.

A few of the professional scribes who produce plain-English documents for communities within the defense establishment are fully conversant with one or more of the tribal dialects. Most, however, are journalists, editors and academics with roots outside of the defense establishment. Thus, they are often as innocent as any other outsider of the actual goings-on in the communities they write about. This innocence, in turn, creates what might be called the "first paradox of defense information" – *the more accessible a document is, the less likely it is to reflect what is really taking place in a particular community.*

In New Guinea, some tribes are more eager than others to greet explorers, explain their customs to anthropologists and tell their stories to journalists. As might be expected, the outside world is more likely to take notice of these tribes and, what is often more important, look at local issues from their points of view. What is true for the tribes of New Guinea is also true for the component communities of the defense establishment. Communities that value engagement

---

[3]Marines refer to their counterpart of an Air Force "FAC" as a "FAC-A," which stands for "forward air controller (airborne)." The Air Force refers to its version of a Marine "FAC" as a "JTAC." Pronounced "jay-tack," this stands for "joint tactical air controller."

with outsiders are more visible than those that shun contact, and are also more likely to influence the way that outsiders think.

The communities within the defense establishment that are most open to outsiders are the ones that lay people are most likely to have heard of. They are the ones that are featured in feature stories, documented in documentaries, celebrated in films, and described in loving detail in Tom Clancy books. They are also the ones that encourage their members to talk to reporters, thereby garnering a lopsided share of press coverage. At the same time, the communities that are less well known are not necessarily those that are cloaked in secrecy. Rather, they have simply developed an institutional culture that places little value on outreach of various kinds.

The fact that some communities are more eager than others to engage the outside world is often a good thing. To begin with, those tribes that are more open to outsiders provide lay people who wish to learn about the defense establishment with a comfortable place to begin their journey. Openness to the outside world, moreover, is not just a virtue where public relations is concerned, but a useful counterweight to the natural tendency of human organizations to focus exclusively on internal matters. At the same time, there is no necessary connection between the amount of publicity a given community enjoys and the role it plays in the grand scheme of things. This, indeed, leads to the "second paradox of defense information" – *the amount of information available about a given community within the defense establishment is independent of its importance*.

Unlike the tribes of New Guinea, the various communities within the defense establishment often engage in advertising of one sort or another. In some cases, such as the glossy pages handed out at military trade shows, the advertising is easily identifiable as such. In other cases, it is harder to distinguish the advertising from information of the other sort. Many of the articles in military trade journals, for example, are based entirely upon information provided by a private company or a government agency. In many cases, moreover, these same organizations provide subsidies of various sorts, whether the purchase of advertising, the provision of office space or donations to related professional associations, for the journals in question.

One of the most interesting things about advertisements related to the world of defense, are peripheral to the words and pictures on the page. For example, the recent proliferation of posters featuring various weapons in Metrorail stations in Washington, D.C. suggests that the makers of such devices are trying to send a message to the thousands of office workers involved in the details of the procurement process. Whether military personnel, civil servants or congressional staffers, none of these people will ever have the power of life or death over the ship, plane or tank in question. Every day, however, one or more of them will choose illustrations for a PowerPoint presentation, type up the agenda items for a meeting or schedule an appointment. Thus we have the "third paradox of defense information" – *some of the most useful information about the defense establishment can be found by studying the context of advertisements.*

The point of this essay is not to make readers cynical. Like a project to study a particular portion of the interior of New Guinea, the task of making sense of a specific community within the defense establishment is difficult, but far from hopeless. Those who undertake such a quest will have to have to learn new dialects, move beyond the information that is most readily available, and master the art of reading between the lines. In doing this, they can comfort themselves with the "fourth paradox of defense information" – *each community within the defense establishment is often as mysterious to members of other such tribes as it is to people from the outside world.*

Essay 4

"Congressional Oversight: Willing and Able or Willing to Enable?"

by Winslow T. Wheeler

Historian Arthur M. Schlesinger Jr. wrote about congressional oversight:

> "The Founding Fathers supposed that the Legislative branch would play its part in preserving the balance of the Constitution through its possession of three vital powers: the power to authorize war; the power of the purse; and the power of investigation."[1]

Congressional investigation, or oversight, is the art of uncovering what is, or has been, going on—why things happened the way they did. With oversight you can –

- understand an issue so legislation can be written with a solution that connects to the nature of the problem, and

- expose mischief in the executive branch, by the opposing party in Congress, or that some other malefactor may be up to, in order to stop or reverse it.

A result of effective oversight might not just be a new law but perhaps an official's resignation, a Justice Department investigation, a program cancellation or the retardation—or advance—of war policy.[2] In successful examples, there is a recurring pattern: facts; that is, previously unknown and important ones, rather than retreads of conventional wisdom, are exposed.

---

[1] Arthur M. Schlesinger Jr. and Roger Burns, eds., *Congress Investigates 1792-1974*, (Chelsea House Publishers, 1975); from "Introduction" by Arthur M. Schlesinger Jr., 11-12.

[2] Find a description of the failure of Congress to exercise meaningful oversight on the vital question of going to war in "The Week of Shame: Congress Wilts as the President Demands an Unclogged Road to War," Winslow T. Wheeler, Center for Defense Information, January 2003, http://www.cdi.org/pdfs/WeekOfShame.pdf.

Mere words, in the form of prognostications at congressional hearings may catch the momentary eye—and the evening news—but their impact on policy, and history, vary from transitory to nonexistent. Beyond that, poorly informed questions, prosecuted ineffectually at a congressional hearing do little more than help us identify which politicians are the lightweights.

I saw exemplar oversight shortly after I started work in 1971 for my first Senate employer, Jacob K. Javits, a liberal Republican from New York. He was a member of the Senate Foreign Relations Committee, then chaired by J. William Fulbright, D – Ark., who held frequent hearings on the disastrous war of that era, Indochina. The hearing I remember was with the secretary of state, William P. Rogers. Fulbright's staff had reported privately to him some U.S. ground combat operations in Laos that violated the Nixon administration's promise to do no such thing. During the hearing, Fulbright repeatedly refuted Rogers' factual assertions about the war, correcting him with information Rogers clearly assumed Fulbright didn't have.

At the time, I was so junior in Javits' office that I had to sit in the public gallery of the hearing room, behind Rogers and his staff. The part I will never forget occurred as Rogers left the room, visibly—but silently—fuming. As he and his unhappy entourage swept past me, one of them growled to an underling, "Find out how those bastards found that out."

Therein find a key initial test for whether any real oversight occurs at a congressional hearing. Are the witnesses leaving smiling, happy to have avoided being put on the hot seat? Clearly, no oversight there. Were they angry and cursing? Well done!

Mixed Record

For recurring negative examples of oversight, I strongly recommend the Senate Armed Services Committee (SASC). From the end of World War II until recent times, the members of the committee and its staff were notorious for being little more than mouthpieces for the Pentagon, being wholly dependent on it for information, advice and direction.[3] There were legitimately autonomous

---

[3] Sometimes, the seeming independence of the SASC has been fraudulent. When he was a member of the committee in the 1950s, future president Lyndon Johnson, D – Texas, issued seemingly revealing reports, but according to one biographer, they were whitewashes and shams. See chapters 13 and 14 of the biography of Johnson's Senate career, Robert Caro, *The Master of the Senate, The Years of Lyndon Johnson, Vol. 3* (Vintage Books, 2002). For a short summary, see also "Cheap Imitator" in chapter 1 of Winslow T. Wheeler and Lawrence J. Korb, *Military Reform: An Uneven History and an Uncertain Future* (Stanford University Press, 2009), 8-10.

members and staffers on the SASC, but they were quite rare. One of the very few I recall were Sen. Harold Hughes, D – Iowa, and a staffer, Charles Stevenson, who were active in the 1970s on exposure of bombing operations in Indochina the Nixon administration said were not occurring. Since that time, the SASC has changed its image to seem a more independent voice, but the absence of any true oversight makes the reality mostly unchanged.

At the SASC's House counterpart, the House Armed Services Committee (HASC), the record is at least mixed. A very notable example of quality oversight on a technically difficult subject was the work of a special subcommittee appointed to investigate the combat record of the M-16 rifle in Vietnam in the 1960s. After several months of investigation—including interviews of troops in the field—the special subcommittee's chairman, Richard Ichord, D – Mo., produced a withering explanation of the jamming failures of the M-16 in Vietnam costing an unknown, but significant, number of U.S. troops their lives. The cause was traced to behavior by Army officials characterized by the official subcommittee report as "unbelievable" and "borders on criminal."[4] In full knowledge of the catastrophic effects on the rifle, the Army changed the ammunition powder, the direct cause of the jamming, and failed to train and equip soldiers and Marines to cope with the ill-effects. When the jamming failures were reported to Army leadership, it failed to take any action until forced to do so by public exposure, and even then the changes made to the rifle failed to address the fundamental problems. (Since that time, M-16 lethality has been further reduced by more Army modifications, and the jamming problem never went entirely away.) The only criticism of the Ichord report that in retrospect seems appropriate is the failure of the subcommittee to call for criminal investigations or resignations.

No Oversight in Sight

Go to any SASC hearing or select any of the archived Web casts at http://armed-services.senate.gov/hearings.cfm. The one I selected in real time occurred on June 15 and 16, 2010. Described at the committee's Web site with the typical oversight title "on the situation in Afghanistan," the hearing came at an especially important point in the controversial war: a major operation in a locality known as Marja was showing signs of falling apart, another for the city of Kandahar had been postponed, and the Afghan President, Hamid Karzai, had been reported to have lost faith in U.S. policy.

---

[4] See pp. 5370-5371 of this exemplar report ("Report of the Special Subcommittee on the M-16 Rifle Program of the Committee on Armed Services," House of Representatives, 90th Congress, First Session, October 19, 1967) in two parts at http://www.vietnam.ttu.edu/star/images/256/2560131001a.pdf and http://www.vietnam.ttu.edu/star/images/256/2560131001b.pdf.

The first sign of non-oversight was the witness list. Invited to testify were the U.S. regional commander for the war, General David H. Petraeus, and a top ranking Pentagon official, Under Secretary for Policy Michele Flournoy. No other witnesses were to be heard; not any authors of independent reports, such as from GAO or a recent, widely reported study about Pakistan's intelligence service undermining the U.S. war effort.[5]

I observed various typical SASC hearing behaviors, including the following.

Dissing the Chairman's Inquiry: SASC Chairman Carl Levin, D – Mich., had a longstanding, public position before the hearing: the Afghan security forces should take on more responsibility for the war. In his opening statement, Levin cited his position at length, and his first question directed at Petraeus was how many Afghan army troops would participate in the upcoming offensive in Kandahar.

Petraeus' answer was short and simple; he didn't know. He made no effort to turn to his staff behind him to give him the data or to—quick—go get it. Instead, he said he would provide the information later, "for the record" of the hearing.

How strange. The committee chairman had a well-known concern; the general and his staff fail to anticipate the obvious inquiry and then basically discount the chairman's inquiry, saying they'll get him something on that later.

Levin showed no sign of being perturbed and asked no follow up question on the matter. Nor did he remind General Petraeus the next day when the hearing continued that he wanted the missing information. The whole exercise seemed to have no point whatsoever.

It would have been simple for Levin and his staff to be much better prepared for his line of questioning. They might have warned Petraeus' staff about the chairman's interest, perhaps even sharing the specific question; so that it was sure to be answered. Had Levin and his staff been really on their toes, they also would have independently researched the answer to their question before the hearing. (Senator Javits once lectured me never to have him ask a question in a hearing I didn't know the answer to.) That way, when they got Petraeus' nothing response, they could say what the data were, point out that the Afghans were not pulling their weight, and drive home the point.

---

[5] Matt Waldman, "The Sun in the Sky: The Relationship between Pakistan's ISI and Afghan Insurgents," Discussion Paper 18, Crisis States Research Center, June 2010, http://www.crisisstates.com/Publications/dp/dp18.htm.

It also would have made quite clear that the chairman was not to be toyed with. This would put the witnesses on warning that they better answer fully and accurately, setting the tone for the rest of the hearing.

That's how it should have been. That's what Fulbright and Ichord would have done. Instead, there occurred a non-exchange of information, and the marker was laid down by Petraeus, not Levin, that he would control what information, if any, was divulged in this hearing.

Was It a Question or a Speech? The third senator to engage General Petraeus was Joe Lieberman, D – Conn. Rather than ask any question, he gave a speech articulating his position on the war as "vital to the national security interests" of the United States. At the end of it all, he gave General Petraeus an opportunity to say pretty much anything he cared to. This is a common tactic at SASC hearings. It is not oversight; it is speech making. Lieberman's exchange of bromides with Petraeus was a classic example.

Do You Want an Answer, Senator? Sens. Mark Udall, D – Colo., Scott Brown, R – Mass., and Kay Hagen, D – N.C., provided other examples of how not to ask questions.

Udall started out saying he would cue up two separate questions and listen to the answers—much like bashful callers on radio talk shows. He got a vague answer to his inquiry about Afghan President Karzai that amounted to little more than Petraeus' saying Karzai had a tough job, and he was told that a study rehashing decades old information about minerals in Afghanistan was the new product of a U.S. bureaucrat who did "phenomenal work." Udall said nothing to indicate he had the slightest disappointment with the useless, even misdirecting, responses.

Brown made it clear Petraeus had nothing to fear by saying he was leaving the hearing soon but wanted to know about contracting and "warlord-ism" in Afghanistan and about Pakistan's counter-Taliban operations (all important issues). Petraeus gave short answers that can be summarized by saying "I'm working on it," and did not even mention that controversial study about Pakistan continuing to help some factions of the Taliban. Incredibly, Brown concluded by thanking Petraeus for his "very thorough answer."[6]

---

[6] Interestingly, the warlord-ism and corruption issue that Senator Brown professed his interest in was the subject of an important report right after the hearing. On June 22, 2010, Congressman John Tierney, D – Mass., released an investigatory report from the House Committee on Oversight and Government Reform on the corruption of warlords and others in Afghanistan in contracting for the transport of U.S. supplies, an operation that had the effect of funneling millions of dollars to the Taliban. Senator Brown, his own staff, and that of the committee clearly had no clue this report was about to be released, let alone of the contents.

Hagen read off from notes presenting a jumble of concerns and ultimately a question about reconciliation between the government of Afghanistan and the Taliban. Petraeus suggested, at least to me, the possibility that it was a question planted by Petraeus or his staff by praising the "nuance" of the question. My notes show no new information transpiring.

And so it went: Mostly bilious questions that weren't really questions and responses that certainly weren't answers. Basically, it was a hearing run by General Petraeus. It wasn't oversight; it was poor theater.

Hearings at the SASC on technical issues, such as a weapon program, are no different. The senators are abysmally informed,[7] don't react when they are being fed pabulum, use the hearing as an opportunity to posture on an issue rather than understand it, and seek out the approbation of the senior military witnesses to show their good standing as pro-defense politicians and, frequently, to ensure DOD's cooperation with the member's pork requests.

Oversight Rules

Oversight is like making your way through a poorly lit maze. Some precautions can help you through.

Precaution #1: The "People Issues" Are the Most Important. Effective oversight is not patty-cake; it will not win you easy friends, fast job offers or fancy retirement parties. However, if the people you investigate try to get you fired, you are probably doing your job well. Expect a stressful experience. If you find that likelihood demoralizing, you are better off doing something else.

Next, consider the member of Congress you are working for. Is he/she in the same political party as the presidential administration you plan to investigate? If so, what is being planned, a whitewash of the political ally's program, or is the member you work for going off the reservation? Just like Senator Fulbright who

---

[7] An example I found particularly sad occurred on March 11, 2010 in a hearing on the F-35 Joint Strike Fighter with Under Secretary for Acquisition, Technology and Logistics Ashton Carter and others. Carter had addressed the unit procurement cost of the aircraft but did not include the development cost (over $50 billion), and he had done so in "base year," not contemporaneous dollars. Using old dollars and incomplete program costs, he was clearly understating the cost of the aircraft. Senator Claire McCaskill, D – Mo., brashly announced in a confident tone of voice that she wanted the costs for the "entire program." But she missed the key points and only asked for the quite minor military construction costs. When the DOD witnesses said they "[didn't] have that number," McCaskill proceeded to lecture them on bringing more complete cost estimates to a hearing, remaining oblivious to what she was missing.

started opposing the Indochina War during his fellow Democrat Lyndon Johnson's administration, there can be intra-party oversight. It can be tricky, but it can and has been done.

Sometimes, oversight is planned to attack political opponents; sometimes it is for reasons of conscience; in either case you have an opportunity to conduct a competent, even fair and objective, investigation. However, if there turns out to be no legitimate basis for pursuing the investigation, you will have to be prepared to tell your political masters that the cupboard of evidence is bare. In the event that they want to proceed nonetheless, you have a legitimate ethical problem on your hands and will have to decide if you want a career as a political hack or a professional.

Next, consider the staff you are working with, and yourself. Is anyone interested in working in a senior position in the Pentagon? In the defense industry? Members of Congress and committee staff directors beware! If you wish to perform defense oversight but if any of your staff is interested in working in the Pentagon or for the defense industry, you should reassign, or better yet fire, them. Their career ambitions will mean they will undermine your investigation by bad mouthing it to others, slow rolling your efforts, declaring the information you want unobtainable, and generally working more for the targets of your investigation than with you.

If it is you who wants a job in the Pentagon or defense industry, you need to resolve your moral quandary. Typically, a Hill staffer will try to have it both ways, but if you have gotten this far in this handbook, you should know you need to make some decisions. If your intent is to be a sell out, everyone will benefit if you do so sooner rather than later.

Genuine staff blood lust for an investigation is necessary but insufficient. Have you or your staff done a successful investigation before? By "successful" I mean, someone was indicted, a program was killed, a manager was fired or resigned, or at least a witness left the hearing very, very unhappy. Certainly, staffers who have limited experience can learn, but it improves your chances for success to have someone around who has already demonstrated skill. The Project on Government Oversight (POGO) offers an entire course on the conduct of oversight. Find a link for this useful opportunity at http://www.pogo.org/cots/.

Last, consider the background expertise you or someone working with you needs. It would be nice to have a retired military pilot, for example, conduct an investigation on a military aircraft, but it is not essential. It is much better to score high on the factors above (willingness and skill) and have a military pilot or aircraft designer source to talk to.

Staffers for the congressional defense committees and for the individual members on those committees frequently score poorly on the most important staff quality measures discussed above. Although they may have technical knowledge, they frequently yearn for jobs in the Pentagon, or as temporary detailees ("military fellows") from the Pentagon, already have them.[8] Many others shun standard oversight ideas, such as inviting witnesses to hearings who have contrary knowledge or points of view; case in point, the Petraeus hearing described above.

In sum, in the defense world on Capitol Hill, you are working in a hostile environment. Hold your enemies close; hold your friends closer.[9]

Precaution #2: The Least Important Issue May Be the Subject of Your Investigation. As they say, DOD is a "target rich environment." Almost any subject you select will command millions or billions of dollars and/or hold American lives at stake. No subject is too mundane. Major scandals have occurred on the subject of travel vouchers, credit cards, and the proverbial DOD hammer and coffeemaker. The key is to follow the matter to its origin. That a hammer cost $400 dollars in the 1980s made for some excited press articles, but explaining how that came to be (and complied with DOD purchasing regulations) reveal important insights about the nature of the Pentagon problem.

Pick a subject that you, better your member of Congress, are interested in, but remember: how effectively you chase down the origin of the problem, rather than how glitzy you describe the horror story you uncover, is the key.

Precaution # 3: Your Evidence Is Your Armor: You must be able to rely on your evidence and, just as importantly, know its limitations. Your enemies will attack you at every opportunity; the slightest chink in your data-armor can cause your downfall, especially if that chink is unknown to you. While evidence that has no compromise in its quality is to be desired; it is also rare.

Approach the presentation of your evidence with the expectation of a hostile audience. You need to convince the unconvinced and the skeptical, not those already on your side. The case you build up to sway the unconvinced will automatically appeal to those inclined to side with you, and it will fortify your relationship with your staff director or member of Congress. When they see you

---

[8] Act with extreme caution toward any of these "military fellows;" expect your every word and action to be reported back to their colleagues and superiors in the Pentagon.
[9] For a description of the hostility toward oversight in Congress' defense committees, see an example of a courageous staffer working for Sen. William V. Roth, R – Del., in Chapter 1 ("One Staffer, Two Senators, and an Investigation") of Winslow T. Wheeler, *The Wastrels of Defense: How Congress Sabotages U.S. Security* (U.S. Naval Institute Press, 2004), 3-8.

have nailed the case, even when you show them the arguments against you, they can gain the confidence that they can press ahead with no ugly surprises.

You may not want to start with collecting documents. You might better start with building human contacts and sources. There are legions of people inside the Pentagon who know more about the subject matter than you will ever hope to; some of them will do everything they can to hinder and oppose you; a small number might help you. Seek the latter out; they are the key to many successful investigations of the Pentagon.

You will need to establish a two-way relationship of trust with these human sources. They will be suspicious of you, and you must be wary of them. Your Pentagon sources will worry that you will carelessly expose them, threatening their jobs or working relationships, and they will worry that you are a wimp who will take their information but do nothing meaningful with it. You should worry that your sources might be advocates and may not give you the full story, and that their own data or analysis might be weak, vulnerable to attack or incomplete. Until you can establish trust, or at least recognize the limitations to the relationship, proceed with all sensors fully on.

With the help and advice of these inside sources, start collecting data. Your sources can give you materials and will tell you what other materials you need. In the 1980s, Congressman Denny Smith, R – Ore., and his staff were led to a testing report on how effective the Navy's Aegis air defense system was, or rather was not, and when the report was finally provided by the Navy, the inside sources helped Congressman Smith identify the missing pages.[10]

You will meet resistance in your data collection. You may have to subpoena documents, and if you can prove you have that power, threaten subpoenas. Many committees require the minority party to co-approve a subpoena. That can be a serious problem, but just as likely is resistance from a senator of your own party who is shilling for the Pentagon. Find out what real powers you have to obtain documents that your target does not want you to get.

If you have no practical subpoena power, you will need to find "work arounds": if the Air Force will not give you that sexy document about the cost of their fighter, perhaps the Navy will; try other sources (perhaps lower in the bureaucracy or in a different but parallel bureaucracy) who are willing to talk; work through GAO which sometimes can be insistent in obtaining documents, or travel to other locations to discuss the same matter with people outside

---

[10] Such examples can be literally unending. One ally, Dina Rasor, started a career by finding "closet patriots" to leak to her unclassified but revealing documents on the M-1 tank, many other weapons, and an unending litany of spare parts horror stories that included the infamous hammer, coffee pot and toilet seat.

Washington. Consider kicking the matter upstairs: Sometimes what you can't get from the assistant secretary, your chairman can get from the secretary. Consider also making a public stink about the document denial.

At some point, you will be offered a compromise: either a chance to review sections of the documents you need or permission for some member of Congress to look at the document, but not you. Resist these smelly deals; they are nothing more than an attempt to feign cooperation while denying you the information you want. Do you really think they will leave key evidence in a redacted document? Also, almost without exception, members of Congress, even staff directors, will be too poorly informed or too busy to understand all the implications of a sensitive document they are shown, usually only for a brief period, rather than to you.

Is there a decent GAO, CBO, CRS or other outside team to work with? Be careful. The quality of the investigators and researchers at these congressional agencies varies greatly. During my nine years at GAO, I found some people there were skilled and aggressive at looking into DOD programs, but I found many who were neither.[11] It is also important to know what different research agencies are good at and what they aren't. While CBO might be helpful on cost issues, it will be less able to help you on technology. Some at GAO might be good on fraud and abuse but not on understanding combat history. The best evidence of the quality of a GAO, CBO or CRS team is its reports. Before you commit to working with any team, read their reports.

Precaution #4: Presentation Matters: There are many ways the results of your investigation can be presented; a hearing is just one. Controlling factors include the nature of your committee and the member you work for. A hearing format may be either a good or a bad idea. Another obvious possibility is the public release of a report in a press conference, or a leak to the press. An exclusive leak to a major newspaper can do the job nicely—if they write the story.

In either case, the release of the material must include the evidence to make it clear to anyone that your case is strong. Don't skimp. At the outset, you do not know what part of your evidence will later prove key in deflating whatever case your enemies try to pump up. There is no such thing as too much support for serious, controversial conclusions and recommendations.

Conversely, it also must be easily digested by short attention span staff and members, and by journalists being hassled by their editors to get a quick story

[11] For a discussion of the quality of GAO's defense work, see chapter 8, Winslow T. Wheeler, "Lapdog and Clouded Lens," *The Wastrels of Defense: How Congress Sabotages U.S. Security* (U.S. Naval Institute Press, 2004).

out. A good, readable executive summary can be crucial, but don't make it so vague that it inadvertently implies the investigation is weak.

Understand your operating environment: the people you are working with and against, the time requirements for getting out your story, and what resources and skills you have—or lack—to get your material into the heads of the people you target to receive it. Your own and others' insights into this environment should lead you to useful conclusions about how to get your message out.[12]

Option B

You've finished your investigation with its hot results, but your member of Congress (or chief of staff) is bailing out. He/she got a call from the nice man at Boeing (or the White House or the Defense Department) and the political decision is to bag your work. Not the first time this has happened; not the last. Politics—the hinted offer of contributions (or a public savaging) or a much desired campaign visit from the POTUS—has overruled your work that might save millions of dollars or provide military personnel more effective training or equipment.[13]

You have options, if you have the stomach for them.

First, make an argument to whoever is blocking your report. Perhaps they are not hard over but want to be assured your work can stand up. Perhaps they have a conscience you can appeal to.

If internal argument doesn't work, you did send that report into your bureaucratic superiors electronically, right? Perhaps a few others' addresses were copied on your message. Did you send it also to officials in the Pentagon,

---

[12] Whatever method you choose to "tell your story," avoid the cheap gimmicks. As the military reform movement was falling apart in the late 1980s, Sen. William Roth, R – Del., and then-Congresswoman Barbara Boxer, D – Calif., held a press conference on the high cost of DOD spare parts. There was not much new to the already thoroughly covered issue. Boxer and Roth decided to jazz it up by decorating a Christmas tree with the various spare parts as ornaments. The glitzy idea was successful in getting the press conference into the news, but the resort to cheap tricks made it clear that military reform on Capitol Hill was out of airspeed, altitude and good ideas.

[13] In my case, it was a member of Congress worried at alienating the Defense Department. In 1997, I had traveled to Fort Irwin, California and came back with information about low military readiness for peacekeeping operations in the Balkans. I had sent my report to Sen. Pete Domenici, R – N.M., but his decision was to do nothing. I nonetheless facilitated a leak to a journalist, and it was published on the front page of the *The Washington Times*; see Rowan Scarborough, "Peacekeeping Puts Drag on Army's Mission," *The Washington Times*, December 23, 1997, 1.

asking them to check over the facts? Perhaps it also went to a few researchers at CRS, GAO and CBO to check on the quality of your analysis.[14]

Get my drift? The electronic age makes it almost impossible to suppress a report once it has passed the electronic portal to a threshold number of people.[15] Wouldn't it be just terrible if it ended up in the hands of some reporter and a front page story was written about your findings?

There will be consequences. For starters, you may lose your job. Politicians and their operatives almost always consider their political comfort more important than your personal fate. There are two possible protections: first, while it is risky, they may see that they are getting some favorable press coverage out of the report; all might be forgiven – perhaps after some finger waving. Or, perhaps you deleted any personal identification from that report that slipped into that reporter's hands and the reporter agreed not to identify you—or anything connected to you—in the article. The report itself should stand on its own legs in terms of data and analysis so mentioning it was a staff report from your office should not be essential.[16]

Of course, losing a job in an office that values politics above content may not be a bad thing. When I lost my job in the Senate Budget Committee because Sen. John McCain, R – Ariz., resented my revelations about his own involvement in and enablement of the congressional pork process, I ended up with a job offer and an invitation to write my first book. Some of these stories can have a happy ending.

---

[14] In distributing these materials, it is essential that they not contain any classified information whatsoever. You have no authority to release it, no matter how legitimately unclassified you might think the material to be and no matter how much you think the public needs to know it.

[15] This phenomenon is not unique to the Internet. In the 1980s, a testing official wrote a devastating report on the performance of an Army air defense system, known as "DIVAD." He distributed 12 copies to his superiors. The head of his office wanted to defend the program and suppress the report; he demanded that all 12 copies be collected and given to him. He received 13. Knowing the gig was up – or rather that the report was being xeroxed – he sent it on to Secretary of Defense Casper Weinberger. When Weinberger learned that at least one congressman also had the report, he cancelled the program.

[16] This became my *modus operandi* when I worked at the Senate Budget Committee. I would periodically permit reports I had written to find their way to the press, ultimately using the pseudonym "Spartacus." Find some details on this behavior, and the consequences, in the preface of *The Wastrels of Defense: How Congress Sabotages U.S. Security* (U.S. Naval Institute Press, 2004).

A Classic Example

In May 1940, before America's entry into World War II, President Roosevelt requested urgent appropriations to pay for America's pre-war build up. As the money flooded into the War and Navy departments, Sen. Harry S. Truman, D – Mo., took it upon himself to visit military facilities to check on how the money was being spent.

Unlike today's regal congressional arrivals at military bases, Truman drove in his own personal car and was not accompanied by a gaggle of military escorts or staffers to arrange his meals and lodging and otherwise pamper him.[17] Truman was horrified at what he found: huge waste everywhere and government officials doing nothing about it. He met privately with President Roosevelt to seek action, but finding no interest in the White House, he delivered a speech in the U.S. Senate chamber and proposed a special committee. The Senate agreed and established a Special Committee to Investigate the National Defense, with Truman as chairman.[18]

Truman time and again invested his own time and energy to understand the issues. He ultimately held 432 public and 300 closed door hearings, conducted hundreds of field trips, and wrote 51 reports. The work addressed aluminum shortages and military construction waste; inefficient production of rubber, aircraft, landing barges, farm machinery and ships; war profiteering; fake inspections of steel plate; the comparative merits of rayon or cotton tire cord; the financing of one U.S. senator's swimming pool and payments to another from defense contractors; and—remarkably for a Democratic-controlled committee—inefficiency induced by labor unions.[19]

Truman and his staff earned a reputation for independence, professionalism and fairness. The chairman did not badger witnesses, and he eschewed topics beyond his proper reach, such as military strategy and tactics; he even kept the committee out of the politically sensitive domain of the location of defense facilities (pork).

Where it did investigate, the committee pulled few punches. Its reports were full of "Truman-esque" barbs; for example

> ... most American pursuit planes were inferior to the best British and the best German pursuit planes... Scarcely a week now goes by without some prominent flyer returning to this country and asking why we can't give the boys better pursuit planes.... the Army should ... give less

[17] David McCullough, *Truman* (Simon & Schuster, 1993), 256.
[18] Schlesinger and Roger Burns, *Congress Investigates*, 337.
[19] Ibid, pp. 335–338.

> attention to concocting publicity blurbs intended to emphasize that poor planes are better than none at all.[20]
>
> So called competitive bidding has often been used as a cover for collusive bidding on Government contracts.[21]
>
> The committee particularly condemns advertising such as the Curtis Helldiver advertising which was intended to give the public the erroneous impression that the Curtis Helldiver was the world's finest dive-bomber and was making a substantial contribution to the war effort when the fact is that no usable plane has yet been produced .... The fact that such advertising was approved by the Navy and was based upon a speech of a Navy Admiral does not justify it.[22]

The overall impact of these and many more frank assertions was not to undermine public confidence in the war effort but to raise it: citizens came to believe the selfish and the inept were being rooted out. One source estimated the committee was responsible for $15 billion in savings, or in modern dollars $270 billion.[23] Others assert that figure is exaggerated, but the savings were "enormous and unprecedented" nonetheless.[24]

Today, the cheap conventional wisdom seems to be that tough oversight over a military at war constitutes questionable patriotism. Truman proves that wrong, even unpatriotic.[25]

Conclusion

Oversight can be difficult, stressful and often thankless, but also rewarding. You will be performing something that is more important than you or the people you work for. (And, you will be pleasantly surprised how many decent people recognize your efforts.) Most importantly, by making Congress and the public aware of important problems and how they came to be, you are performing one

---

[20] *Congressional Record*, January 15, 1942. The Xerox of these pages provided by the Library of Congress to the author did not include the page numbers.
[21] *Congressional Record*, March 4, 1944.
[22] *Congressional Record*, July 10, 1943.
[23] Schlesinger and Roger Burns, *Congress Investigates*, p. 338.
[24] David McCullough, *Truman* (Simon & Schuster, 1993), 288. On the other hand, Senator Truman's work was not without compromises. He did not look into racial discrimination in hiring at defense plants and segregation in the military services.
[25] For the details of how Truman did what he did, see the David McCullough biography and *Congress Investigates 1792-1974* by Arthur M. Schlesinger Jr. and Roger Burns, referred to above.

of the most important functions of government that our Constitution calls for—and needs for a democracy to survive.

Essay 5

"Careerism"

by G.I. Wilson

This essay attempts to make it easier for you to identify the quality and character of military officers and civilian bureaucrats you meet, socialize and work with - to increase your awareness and recognition of careerism and its consequences. As Americans, we all must exercise more care and caution in our appraisal of our senior military officers and the Washington "suits" that exert dominating influence on the cost of defense and the conduct of American national security policy.

The Department of Defense (DOD) that I have observed all too closely for over three decades is an overgrown bureaucracy committed to standing still for, if not actively promoting, poorly conceived policy agendas and hardware programs funded and supported by Congress. Coupled to that is the task of attracting the blind loyalty of senior military and civilian actors on the Washington, D.C. stage. For the careerists in America's national security apparatus, it is all about awarding contracts and personal advancement, not winning wars.

Careerists serve for all the wrong reasons. They weaken national defense, rob the military of its warrior ethos and drive away the very highly principled mavericks that we need to reverse the decay. This can only be remedied by rekindling the time honored principles of military service (i.e. duty, honor, country) among both officers and civilians.

What Is Careerism?

In the DOD today, standard bureaucratic behavior is focused on conniving with politically focused congressional advocates and their counterparts in industry and think tanks to advance selected hardware and policy agendas. Once the careerist generals, admirals, colonels and captains exit active military service, they perpetuate their inside baseball by re-materializing as government appointees, political candidates, DOD contractor shills, so-called Pentagon "mentors," and network talking heads. All are raking in money, peddling influence, exerting pressure for vested interests, all the while collecting retired pay, healthcare, commissary privileges and more at taxpayer expense.

For example, Gen. Jim Jones, U.S. Marine Corps, ret., occupied a big chair in the White House as the president's national security advisor. Adm. Joe Sestak, U.S. Navy, ret., went to Congress as a member of the House of Representatives seeking promotion to the U.S. Senate. Lt. Gen. Karl W. Eikenberry, U.S. Army, ret., is the U.S. ambassador to Afghanistan. Many others dot the boards of the big defense contractors. As author Bob Woodward points out in *The War Within*[1], many of the uniform-to-suits careerists made themselves cozy with political circles in Washington, D.C. in ways and to a degree that did not exist before 2001. As for the senior careerists in the ranks of the civilian bureaucracy, there is a similar variation of take-this-job-and-flip-it among public, academic and private sector positions. While it's distasteful observing this in civilian quarters, it is the "self-fixation" of our top military leadership that this author finds most disturbing.[2]

The Problem as Described by Others

What is wrong with retired officers populating civilian government offices, industry and politics?

Author Edward N. Luttwak explains that it means a lifelong path of political correctness, playing it safe, making only decisions that create no waves, or – better yet – waves that promote the selected agenda. Worst of all, careerists leverage the bureaucracies in DOD and Congress to dilute any personal accountability and responsibility - the very essence of careerism. Luttwak warns "If careerism becomes the general attitude, the very basis of leadership is destroyed."[3] That era of pseudo-leadership is upon us.

Careerism is also artfully described by Robert Coram and Col. John Boyd of the U.S. Air Force. The careerist's singular aspiration is "the desire to be, rather than the desire to do. It is the desire to have rank, rather than use it; the pursuit

---

[1] Bob Woodward, *The War Within: A Secret White House History 2006-2008* (Simon & Schuster, 2009).

[2] Journalist Bryan Bender wrote an extraordinary analysis of this behavior in the December 26, 2010 *Boston Globe;* see "From the Pentagon to the Private Sector" and related materials at http://www.boston.com/news/nation/washington/articles/2010/12/26/defense_firms_lure_retired_generals/

[3] Edward N. Luttwak, *The Pentagon and the Art of War* (New York: Simon and Schuster, 1984).

of promotion without a clear sense of what to do with a higher rank once one has attained it."[4]

The etiology of careerism stems from a shift in the basic values within the officer corps as described by Samuel P. Huntington in his classic work *The Soldier and the State: The Theory and Politics of Civil-Military Relations*.[5] Huntington contends the most important feature that distinguishes military personnel from all others is the view that the military is truly a "higher calling" in the service of one's country.

Today, this is no longer the case. Morris Janowitz observed in *The Professional Soldier: A Social and Political Portrait*:

> Those who see the military profession as a calling or a unique profession are outnumbered by a greater concentration of individuals for whom the military is just another job . . . . For a sizable majority - about 20 percent, or about one out of every five - no motive [for joining the military] could be discerned, except that the military was a job.[6]

Maj. Michael L. Mosier posits in "Getting a Grip on Careerism" in *Airpower Journal* how military sociologists theorize that the idea of a higher calling has diminished as institutional values deteriorate.[7] While institutional values deteriorate, careerists exhibit traits of psychopathy replacing the higher calling with ambitions of personal gain and unaccountability.

Babiak and Hare's *Snakes in Suits*, a book about psychopaths in the workplace, may seem foreign when juxtaposed with national security, but is instructive in the recognition of character traits the careerists exhibit and the wreckage they leave behind. (The writer is not suggesting that all careerists are psychopaths; however, the behavior of both has much in common.)

Consider the behavior of psychopaths described by Babiak and Hare: Glibness, superficial charm, grandiose sense of self-worth, deceitful, cunning, manipulative, lacks remorse, callous, lacks empathy, does not accept

---

[4] Robert Coram, *Boyd: The Fighter Pilot Who Changed the Art of War* (Boston: Little, Brown and Company, 2002).

[5] Samuel P. Huntington, *The Soldier and the State: The Theory and Politics of Civil-Military Relations* (Cambridge: The Belknap Press of Harvard University Press, 1957).

[6] Morris Janowitz, *The Professional Soldier: A Social and Political Portrait* (New York: Free Press, 1971).

[7] Maj. Michael L. Mosier, "Getting a Grip on Careerism," *Airpower Journal* 2, no. 2 (Summer 1988): 52-60.

responsibility for own actions, and impulsiveness.[8] Look for these behavioral markers in careerists and what psychologists call the "paradox of power."

Jonah Lehrer writes about the "paradox of power" in *The Wall Street Journal* contending that the very traits that help leaders accumulate power and influence in the first place (being polite, honest, outgoing) all but disappear once power and authority are achieved. Positive leadership traits are replaced with impulsiveness, recklessness and rudeness. Lehrer further notes that authority coupled with the power paradox leads to flawed cognitive processes that in turn "distort the ability to evaluate information and make complex decisions."[9]

As one who has worked in and around the Pentagon bureaucracy for a few decades, other characteristics come to mind. In addition to placing one's self in a position of accelerating personal gain, careerists also collect accoutrements of rank and position, perks and lists of biographical achievements, defined as positions, ranks and titles held. It is not about what they achieved but rather the positions and titles they held.

It is appalling that so many senior officers think that the military is all about getting promoted and accumulating as many signs of rank and status as possible, completed with a host of perks. What is lost on careerists is that they are getting the opportunity to actually do things that most people only dream of, or get to see just in the movies.

They are so prevalent because bureaucracies are in effect designed by and for careerists propagated by reams of regulations and layers of superfluous commands. Bureaucracies give careerists a place "to be somebody" rather than an opportunity to do something. They are promoted because of a zero defect record of playing it safe, making no controversial decisions and requiring others to do the same.

Recognizing Careerists

Careerists in both uniforms and suits thrive on hardware programs. It is not a matter of whether a weapon system works but whether it survives. One might point to the failed programs like the A-12 bomber or the Sgt. York "DIVAD" gun which saw billions wasted before they were cancelled. But look more skeptically at the programs that survive, even prosper, that are irrelevant to the

---

[8] Paul Babiak and Robert D. Hare, *Snakes In Suits* (New York: Collins Business, 2006), 27.

[9] Jonah Lehrer, "The Power Trip," *The Wall Street Journal*, August 14-15, 2010.

wars we fight, double in cost (or more), are delivered years late and break promise after promise for performance.

Even for the so-called successful programs, the improved performance is never commensurate with the increase in cost. What manager among the orchards of low hanging fruit of Pentagon procurement fiascos has been held accountable? What senior DOD acquisition "Czar" has not found himself a huge pay raise from industry upon retirement? Congress and DOD often reward poor program performance and cost escalation. In 2010, Defense Secretary Gates replaced the general in charge of the Joint Strike Fighter program, but the action was a remarkable exception, and nothing fundamental to the program's problems was changed.

Recognizing the Ego Factor

The careerists are not interested in fostering people and ideas or developing good personnel and education programs. The rewards are in hardware issues, not people issues - except that one human factor does predominate: self.

The Washington Post wrote a review of Gen. Wesley Clark, U.S. Army, who was relieved of command in Europe in 2000 shortly after the ineffectual military campaign he commanded against Serbia in 1999. (Not long thereafter he immersed himself in presidential campaign politics.) The article revealed much about the man's careerism and its characteristics. The reporter for *The Washington Post* explains with details the animus against Clark: His leadership was "undercut by his relentless need to be front and center, to always make it all about him winning -- rather than the mission."[10]

Clark's deep infatuation with the word "I," which runs through the veins of all careerists, was evident in his own explanation to the reporter:

> "How do you think I could have succeeded in the military if everybody didn't like me? It's impossible," he said. "Do you realize I was the first person promoted to full colonel in my entire year group of 2,000 officers? I was the only one selected. Do you realize that? . . . Do you realize I was the only one of my West Point class picked to command a brigade when I was picked? . . . I was the first person picked for

---

[10] Lois Romano, "A Hero To Some; To Others, Headstrong," *The Washington Post*, October 19, 2003.

> brigadier general. You have to balance this out. . . . A lot of people love me."[11]

If Clark blames himself at all for the abrupt ending of his career after 34 years in the Army, he has never let on. More than one friend has quoted him, when trying to comprehend his forced retirement, as saying plaintively, "But we won the war..."

Without question Clark, like most careerists, has little love for subordinates, peers and others whom he sees as impediments to his career. *The Post* reported "In an institution filled with ambitious men, some viewed Clark as over the top, someone who would do or say anything to get ahead -- and get his way."

Placing self above the interests of one's military service, DOD, and even national security is *de rigueur*. The *Taipei Times* of Sept. 9, 2010, wrote of retired U.S. Navy Adm. William Owens the following:

> Retired US Admiral William Owens — the former vice chairman of the Joint Chiefs of Staff who wants to end arms sales to Taiwan — is now aiding an effort by China's Huawei Technologies to supply equipment to Sprint Nextel and operate in the US.
>
> A team of eight US senators has written to the administration of US President Barack Obama warning that the move by Huawei could "undermine US national security."
>
> A national carrier in the US servicing 41.8 million customers at the end of the second quarter, Sprint Nextel is also a supplier to the Pentagon and US law enforcement agencies.

And later,

> If our electronics are compromised, we are cooked," [China expert Arthur] Waldron said in his e-mail sent to a wide circle of China watchers.
>
> "Who is to say that subsystems bought from China will not have back doors and hidden links to their suppliers? We would be mad to think otherwise. The Chinese are not stupid," he wrote.[12]

---

[11] Ibid.

Recognizing the Silence of Careerism

The same careerist system rewards those who ignore hardware but promote, or fail to stand up against, gigantic policy mistakes. Ambassador Paul Bremmer, who was awarded the presidential Medal of Freedom, insisted on the disbanding of the Iraqi army in May 2003. This put an estimated 350,000 to 400,000 Iraqi soldiers out of work, and available to help foment the violence that followed.

Many serving officers and retirees are not forgetting that when senior commanding generals of America's expeditionary ground forces assembled in Baghdad in May 2003 to hear Ambassador Bremer announce the decision to dismantle the Iraqi state, army and police and occupy much of Arab Iraq with U.S. and British forces, not a single general officer raised any objection.[13]

It is impossible to know whether the refusal of general officers commanding American forces in the field to implement such a misguided and disastrous policy would have allowed American forces to avoid the expensive occupation of Iraq. Speaking out or retiring immediately certainly would have given officials in the government an opportunity to consider places a thousand times more important than Diyala or al-Anbar, starting with the United States itself."[14]

The apologists for this behavior deceptively ascribed their ruthless climb of the Pentagon ladder as an artifact of doing the right thing. But it is actually a lack of professionalism and an abandonment of the principles of military service. The conflicts in Iraq and Afghanistan provide the most painful recent examples. They have severely tested and frequently compromised the U.S. officer corps' traditional values of duty, honor and country. This is obvious in the selective

---

[12]William Lowther, "Owens' links to PRC firm ring alarm," *Taipei Times*, September 9, 2009, http://www.taipeitimes.com/News/taiwan/archives/2010/09/09/2003482460.

[13] David Phillips, *Losing Iraq: Inside the Postwar Reconstruction Fiasco* (New York, NY: Perseus Books, 2005), 145. As if to reinforce his support for disastrous policies, General Petraeus, who was present for Bremer's announcement in May 2003, said nothing in response. Instead he insisted in an interview with the Iraq Study Group on May 18, 2006: "'US Strategy over the last 18 months has been sound. The ongoing violence had made the mission more difficult. Nonetheless, no alternative strategy is better." Petraeus added the United States had "terrific people" assigned to the war, endorsing General Casey and Ambassador Khalilzad and noted, "I would not break up the team of military and civilian leaders currently in Iraq." See Bob Woodward's book, *The War Within: A Secret White House History 2006-2008*, 44.

[14] Edward Luttwak, "Errors of Backsight Forethought," *Politics*, October 2009, 31.

careerist- and agenda-ridden assertions to portray a false picture of events to the American public about the Iraq and Afghanistan wars. Recent examples from every level of command are:

- Americans were told Iraq was invaded to locate and destroy weapons of mass destruction. It was a lie.
- Americans were told former National Football League star Army Ranger Sgt. Pat Tillman died fighting the enemy. It was a lie.
- Americans were told Army Spc. Jessica Lynch fired her M16 rifle until she ran out of bullets and was captured. It was a lie.
- Americans were told repeatedly the rebellion against our military presence in Iraq was defeated and "security was improving." It was protracted lying punctuated by a daily diet of exploding bombs and mutilated bodies until massive cash payments to the Sunni Arab opponents bought cooperation.
- Despite numerous classified and unclassified accounts of brutality meted out to prisoners of war and the civilian population by U.S. forces in Iraq and Afghanistan - reports that describe the chain of command as aware of the abuses but routinely ignoring or covering them up - not a single general officer was called to account.[15]
- In 2010, Americans are told Iraq is a "democracy," when in reality, Iraq is mired in corruption and violence,[16] its oil is in Chinese hands,[17] and Iran, not the United States influences Iraq's political destiny.[18]

---

[15] Adam Zagorin, "Pattern of Abuse: A decorated Army officer reveals new allegations of detainee mistreatment in Iraq and Afghanistan. Did the military ignore his charges?" *Time*, September 23, 2005, 32.

[16] Jane Arraf, "Iraq bomb before election has some fearing new civil war," *The Christian Science Monitor*, February 18, 2010, 1. Also see Doug Bandow, "Bombs Away: Conservatives Embrace War," CampaignforLiberty.com, February 10, 2010. Also, see Scott Peterson and Howard LaFranchi, "Iran shifts attention to brokering peace in Iran. Details from a secret meeting between top Iranian and Iraqi officials signal Iran's aim to 'stop arming' militias," *The Christian Science Monitor*, May 14, 2008, 1.

[17] Kyle B. Stelma, "Report: Private Foreign Direct Investment in Iraq," (Washington, D.C. and Dubai: Dunia LLC, 2009), 13-14.

[18] David Phillips, *Losing Iraq: Inside the Postwar Reconstruction Fiasco* (New York: Perseus Book), 145.

One can go on, especially now about Afghanistan, but surely the point is made: as the American people are told the conjured tales of the policy advocates, the senior military command stays silent; in fact, some assist, even fabricate, deceptive rationalization further underwriting deafening silence.

Effects

President Eisenhower's worst nightmare described in his January 1961 farewell address has become fulfilled. Today's consolidated defense industries have become inseparable from the government and hold political careers in the U.S. Senate and the House of Representatives at risk if sufficient tax dollars are not committed to the industries' expensive defense products.[19] That the politicians succumb, holding their political well-being above the merits of any weapons debate, is the very definition of careerism. Unless and until the politicians realize their political fate hinges on a broader perspective, their votes on defense issues will be driven by their narrowly perceived short-term interest, mostly "pork" and campaign contributions.

The "revolving door" enriches civilian executives in the defense industry, and its supporting consulting businesses, for periodic service in the Department of Defense, and it rewards retired generals and admirals for their access to the men and women they left behind in the Pentagon and not coincidentally promoted to flag rank. Rewards are particularly plentiful for the three- and four-star officers who supported and defended expensive defense programs even when the usefulness of the programs was doubted inside their own service bureaucracies, among other places.[20]

Consequently, it's no surprise that federal auditors, poring over the Defense Department's conflicting financial statements, missing data and accounting discrepancies, are unable to provide an accurate accounting of the Defense Department's books.[21] According to a July 8, 2004 report by the Government

---

[19] Micah L. Sifry and Nancy Waltzman, *Is that a Politician in your Pocket? Washington on $2 Million A Day* (Hoboken, NJ: John Wiley & Sons, 2004), 6-9.

[20] Ann Roosevelt, "Future Combat System Is 'Real,' Army Will Work to 'Protect' It, Top Leaders Say," *Defense Daily*, October 10, 2007, 11. "'I will tell you that it's real,' Army Chief of Staff Gen. George Casey said at the same event." Two years later Casey was ordered by Secretary Gates to cancel FCS.

[21] Rowan Scarborough, "U.S. Auditors Homed In on Hillah Contracts," *The Washington Times*, November 28, 2005, 4. Also, see Stephen Glain, "Cashing In on America's Wars: Waste, Fraud, and a Cast of Thousands," *The National*, July 1, 2009, 2., and Paul B. Farrell, "America's Outrageous War Economy! Pentagon can't find $2.3 trillion, wasting trillions on 'national defense," *Market Watch*, August 28, 2008, 13.

Accountability Office, the generals in U.S. Central Command and Washington, D.C. lost $1.2 billion worth of war materiel shipped to Iraq for the campaign to remove Saddam Hussein from power.[22] More recently, a congressional staff report found aid to Afghanistan ending up in the hands of the Taliban.[23] This sort of thing would almost be funny, in an insane sort of way, if poor senior leadership did not result in the loss of American life in uniform, undermine American strategic interests abroad, drain the United States Treasury of its hard-earned tax dollars, and erode the economic well-being of the American people the nation's flag officers are sworn to defend.

Perhaps, the lack of accountability explains why supposedly objective, retired military officers retained as analysts by national television networks have little incentive to jeopardize their lucrative contracts with the political and industrial elites to tell the American people the hard facts about events in Iraq or Afghanistan? Nurturing the Pentagon money flow and the domestic political environment that supports it while influencing their chosen successors—often their former aides—to keep the money spigots open profoundly changes the message the retired generals and colonels send to the listening audience. [24]

These behaviors help reinforce the myth that only generals and admirals can or should formulate the fundamental principles governing the application of American military power, or even military doctrine.[25] Today, this myth has transformed the president, as well as members of the House and the Senate, into

---

[22] David Wood, "Auditors Despair over Pentagon's Books," *San Diego Union-Tribune*, July 21, 2004, 1.

[23] The House Committee on Oversight and Government Reform released an investigation on contractor corruption in Afghanistan in June 2010; find it at http://oversight.house.gov/images/stories/subcommittees/NS_Subcommittee/6.22.10_HNT_HEARING/Warlord_Inc_compress.pdf.

[24] For example, see Tom Vanden Brook, Ken Dilanian and Ray Locker, "Retired military officers cash in as well-paid consultants Netvibes," *USA Today*, November 17, 2009, 1. See also Janine R. Wedel, *Shadow Elite: How the World's New Power Brokers Undermine Democracy, Government and the Free Market* (New York: Basic Books, 2009), and David E. Johnson, "Modem U.S. Civil-Military Relations. Wielding the Terrible Swift Sword," McNair Paper 57, July 1997.

[25] Bill Roggio, "McChrystal to resign if not given resources for Afghanistan," *Threat Matrix*, September 21, 2009. Roggio writes: "Within 24 hours of the leak of the Afghanistan assessment to *The Washington Post*, General Stanley McChrystal team fired its second shot across the bow of the Obama administration. According to McClatchy, military officers close to General McChrystal said he is prepared to resign if he isn't given sufficient resources (read "troops") to implement a change of direction in Afghanistan."

doormats for the four-stars.[26] Secretary of Defense Gates and the Army and Marine Corps four-stars in U.S. Central Command (CENTCOM) currently wield more influence over U.S. defense and foreign policy than any senator or congressman, and almost no one in the mainstream media is willing to challenge anything they say or do.[27]

Renewed enthusiasm in the four-star ranks for pursuit of the presidency is surely also related to these trends. It's no secret that a four-star general who transforms himself into a political figure while still in uniform with the aid of political allies in the press and Congress can be so powerful the president may be reluctant to publicly oppose him.[28] After all, members of Congress are always willing to cultivate outspoken four-star generals for narrow partisan advantage.[29] Gen. David Petraeus, the current CENTCOM commander, is the latest in the succession of Army four-stars (including former NATO Commanders Alexander Haig and Wesley Clark) who clearly harbors, despite denials, aspirations to be president.

It is against this backdrop of tumultuous change in civil-military relations since Eisenhower left office that officers coming to Washington, D.C. for the first time - in many instances from arduous duty as company, battalion or brigade commanders in Iraq and Afghanistan - must be viewed. These are the officers that members of Congress and their staffs are likely to meet, and it is from their ranks that will spring the next generation of flag officers. Understanding what makes these officers tick is the real challenge.

Understanding Military Officers

It's impossible to talk about officers in the armed forces without some mention of demographics. As in the past, the overwhelming majority of officers (roughly

---

[26] For example, see Robert Dreyfuss, "The Generals' Revolt. As Obama rethinks America's failed strategy in Afghanistan, he faces two insurgencies: the Taliban and the Pentagon," Rollingstone.com, October 28, 2009.

[27] Bob Woodward and Gordon M. Goldstein, "The Anguish of Decision," *The Washington Post*, October 18, 2009. "Bundy said that Johnson viewed the general as though he were a powerful constituency wielding vital legislative votes."

[28] "General David Petraeus tipped as Republican 2012 presidential candidate," *The Daily Telegraph*, March 19, 2010.

[29] James Parco and Dave Levy, *Attitudes Aren't Free: Thinking Deeply About Diversity in the U.S. Armed Forces* (Maxwell Air Force Base, Alabama: Air University Press, 2010), 408.

75 percent) are of European ancestry. However, regardless of their ethnic origin, American officers are more likely to be from high-income families and they are on average better educated than most American citizens.[30] This demographic profile is consistent with historic data in all, but one way. Today's officers are more religious than their predecessors were 20 or 30 years ago,[31] and they've grown up inside a military bureaucracy that differs in important ways from the Reagan-era armed forces.[32]

There are other factors as well. Today, the new paradigm of warfare (counterinsurgency) creates bureaucratic power bases and careerists that derive their relevance from the currently accepted view of war. Few, if any, military officers rose to prominence in the aftermath of the Vietnam War by arguing for an institutional doctrine that addressed the complexities of limited wars. Today, just about no one will rise through the ranks by raising issues about the U.S. armed forces' ironically new exclusive strategic focus on counterinsurgency. The overemphasis on counterinsurgency must be countered by candid debate and coming to grips with fourth generation warfare - the legacy of failed states and hybrid threats.

The tendency inside the peacetime military to advance officers who tell the boss what he wants to hear is well known;[33] being candid is not career enhancing. This chronic lack of professional candor is now a pervasive facet of political correctness and careerism that supports a new doctrinal orthodoxy inside DOD. That new orthodoxy is a doctrine based in part on a popular journalistic narrative that is deeply flawed but coincides with the careerist *modus operandi* of going along to get along. In practice, the advocates of this doctrinal orthodoxy are not telling U.S. ground forces to adapt to future strategic conditions and global hybrid threats. They are instead telling American forces to train and equip almost exclusively for future unwanted occupations inside the Islamic world.

---

[30] Alfred Vagts, *A History of Militarism: Civilian and Military* (New York, NY: The Free Press, 1959), 492.

[31] Barry Fagin and Lt. Col. James Parco, U.S. Air Force, "A question of faith. Religious bias and coercion undermine military leadership and trust," *Armed Forces Journal*, January 2008, 40. "US military accused of harboring fundamentalism," *AFP*, February 13, 2008, 1. Also see: Headquarters, United States Air Force, "The Report of the Headquarters Review Group Concerning the Religious Climate at the U.S. Air Force Academy," June 22, 2005.

[32] Robert Maginnis, "Distrust Corroding the Military," *The Washington Times*, March 2, 2000, 11.

[33] Leonard Wong, *Stifled Innovation? Developing Tomorrow's Leaders Today* (Strategic Studies Institute, U.S. Army War College, April 2002).

Unfortunately, the officers advocating doctrinal orthodoxy and persistent warfare inside the Islamic world are as career-minded and oppressive as those who maintained the fiction that Operation Desert Storm validated warmed over "Blitzkrieg theory" in the form of air-land battle doctrine in 1991.[34] The use of the term "counterinsurgency" to describe conflicts inside the Muslim world creates the illusion the United States has "discovered" a military solution to societal misery. This assertion is untrue, and officers who've served for years in places where no sane American would voluntarily spend two minutes will make these points in private if asked.[35]

Many officers today think America's national security demands armed forces organized around the capability to fight enemies with the capability to fight back - enemies that look like our own conventional forces and are not optimized for counterinsurgency, or even split down the middle that try to do both.[36] A major with two tours in Iraq and one in Afghanistan summed up the problem that weighs heavily on the minds of many officers in the Army, Air Force, Navy and Marines:

> If we were to fight against someone who was capable and at least marginally equipped, we could, for the first time since the Korean War or World War II, find ourselves fighting on someone else's time schedule and initiative. No one in the force today knows what it is to fight on someone else's clock. If we were hit and hit hard during a build-up, if we faced a capable anti-air threat that knocked a few aircraft, manned or unmanned, out of the sky, against a naval threat that could actually threaten our surface combatants in coastal waters, or that had a ground force that could give battle and launch surprise attacks of their own, our collective psyche's would be shocked, and our forces paralyzed.[37]

---

[34] This is explained in "Operation Iraqi Freedom: Third Infantry Division (mechanized) After Action Report, Final Draft," May 12, 2003. The document is not available online but is in the author's files.

[35] They might also reluctantly utter the words "Fourth Generation Warfare" to explain how failed-state provocateurs, non-state actors and terrorists prescribe that everything goes in war, including not playing by the rules of nation states. See William S. Lind, Keith Nightengale, John F. Schmitt, Joseph W. Sutton and Gary I. Wilson, "The Changing Face of War: Into the Fourth Generation," *Marine Corps Gazette*, October 1989.

[36] Col. Gian Gentile, U.S. Army, "The Imperative for an American General Purpose Army That Can Fight," *Orbis* (Summer 2009): 457.

[37] The officer asked not to be identified. He is now a serving lieutenant colonel.

In sum, our armed forces today are tasked to fight occupational wars they cannot win and they are unprepared for the enemies we claim to be best suited for. That the voices you can faintly hear expressing concern about this (and the assertion that it is not a hardware acquisition question – i.e. a money-making issue) come from the middle of and beneath the officer corps shows how vacant the careerist minds at the top have become.

The Officer Corps in the Balance

In years past, it was easy to identify officers who spent their time checking with superiors or peers concerning whether or not to act. These types seldom pursued what was right. They were simply "staying in their lane," as the saying goes. Officers with the moral courage to take a stand on the grounds that it was in the interest of the American people, even when it might contradict the service's bureaucratic guidelines, were not easy to find, but not uncommon. Today, officers with these attributes still exist, but they are very hard to find. Officers who do so now must be extremely clever, as well as extraordinarily courageous. The erosion that caused this change is an important change that outsiders, including journalists and Hill staffers, must grasp and appreciate.

Officers' disenchantment with the nation's focus on hostile occupations and armed nation-building is matched by a growing lack of confidence in, and recognition of careerism among, the field-grade officers, i.e., colonels and generals, but also those senior enlisted who have opted for careerism - aping their officers.

My personal experience and recent surveys indicate that junior officers in the U.S. Army (and Marine Corps) feel a lot of dissatisfaction with the quality of senior leadership. This "disconnect" between junior officers, and their commanders, has been around for more than a decade. It's gotten worse with a war on, because, unlike past wars, there has not been widespread removal of battalion and brigade commanders who did not perform well. In World War II and Korea, it was common for commanders who did not deliver, to be replaced. With a war going on now and junior officers facing life and death situations because their commanders were not being aggressive or innovative enough, many have been leaving the service.[38]

Lt. Col. Peter Kilner, U.S. Army, returned in 2009 from two months in Iraq where he interviewed young Army officers for a research project. His observation reinforces the comments above: "There is enormous pride among

---

[38] "The Boss Is An Idiot And Is Getting Us Killed," StrategyPage.com, December 23, 2009.

young officers in their units and in each other, but I see strong evidence that they are rapidly losing faith in the Army and the country's political leadership."[39] Careerism and political correctness in all the services may be taking a much greater toll (although a somewhat different one) on our personnel than the enemy in Iraq and Afghanistan.

Concluding Thoughts

The U.S. military is not led by a Centurion or Spartan class of hardened professionals. Perhaps it should be. The leadership of the armed forces looks bleak, save for a very few. The outliers among senior officers are those who are willing to take unpopular positions for the troops' or nation's benefit (not for their own benefit and career enhancement) on politically charged issues. For example, Generals Conway and Amos articulate opposition inside the Marine Corps to the repeal of the "Don't Ask; Don't Tell" policy regarding gay and lesbian service members (DADT), reflecting a sentiment in the Corps' ranks. Whether or not one agrees or disagrees with DADT, is not the issue. The point is Generals Conway and Amos have the moral courage to state their position as unpopular as it may be in some politically correct circles. This writer submits too many, unlike these generals, would rather go along to get along.

For the moment, U.S. military culture and the essence of conducting warfare within clearly defined Constitutional and sensible strategic parameters are insidiously perverted by domestic political interests, political correctness and political constituencies inside the senior ranks of America's military establishment fused to the generals' and admirals' unabashed careerism.

The questions members of Congress and journalists should ask are the questions on the minds of many officers in the armed forces regarding these issues: Are the four-star generals and admirals merely military "caretakers" for the assigned mission without taking moral or professional responsibility for the assignment to which American military power is committed? Are conflicts with Islamic groups that have no armies, no air defenses and no air forces yet another avenue for generals, admirals and colonels to pursue selfish ends?

Lt. Col. Paul Yingling writes about the failure to resist utterly stupid and self-defeating policies conceived in Washington, D.C. Yingling contends that this failure is not the result of "individual failures, but rather a crisis of an entire

---

[39] Greg Jaffe, "Critiques of Iraq War Reveal Rifts among Army Officers. Colonel's Essay Draws Rebuttal from General; Captains Losing Faith," The *Wall Street Journal*, June 29, 2007.

institution."[40] America's generals and admirals have failed to prepare our armed forces for war, yet they advise civilian authorities on the application of force to achieve the aims of policy.[41]

Meanwhile America's generals, colonels, admirals and captains blinded by the illusion of bureaucratic power, mimic the behaviors of the politicians, managers and policy advocates. Individuals preoccupied with their own internal goals are blind to what is happening around them: "Being in a position of power makes people feel they can do no wrong."[42] As a result of this intoxication with power, careerists unwittingly (and wittingly) underwrite a defense-industrial-congressional complex where the primary purpose is awarding contracts and shoveling power, perks and money in disparate forms, rather than winning wars.

How do we fix this? Part of the answer is military reform ushered in by drastic budget cuts to hardware programs (which are addressed in the essays addressing budget, acquisition and weapons in this handbook). Col. Michael Wyly, U.S. Marine Corps, ret., who is known to many of the authors of this handbook and held in high respect, seeks a culture where a warrior class of "mavericks" is accepted and those who place themselves above the time-honored principles of military service (duty, honor, country) find themselves on the outside looking in. Wyly observes of the consummate Pentagon maverick, Col. John Boyd (discussed throughout this handbook):

> Yet it is unfortunate that we have to think of him as a maverick. He should have been the norm: an independent thinker who did his own research on a daily basis and espoused his views regardless of convention because he had the courage to do so. Courage is a virtue. In the military profession, courage tops the list of virtues required and demanded. My experiences in combat demonstrated that you can't have the physical kind of courage without the moral kind. Officers with Boyd's degree of moral courage need to be the norm, not the mavericks. Another way of putting it is that we all need to have the courage to be

---

[40] Lt. Col. Paul Yingling, "A Failure in Generalship," *Armed Forces Journal*, May 2007.

[41] Statement of Dr. Janet Breslin-Smith, House Armed Services Committee Subcommittee on Investigations and Oversight, May 20, 2009.

[42] Jason Zweig, *The Wall Street Journal*, October 16-17, 2010.

mavericks when institutional thought stagnates. But we have a responsibility not to let it stagnate.[43]

[43] Col. Michael D. Wyly, "In Praise of Mavericks," *Armed Forces Journal*.

Essay 6

"Confused Alarms of Struggle and Flight: A Primer for Assessing Defense Strategy in the post-Iraq World"

by Chet Richards

War no longer exists. Confrontation, conflict, and combat undoubtedly exist all round the world ... and states still have armed forces which they use as a symbol of power. None the less, war as cognitively known to most non-combatants, war as battle in a field between men and machinery, war as a massive deciding event in a dispute in international affairs: such war no longer exists.[1]

A National Defense Strategy for the United States

Suppose someone asks you to assess a national defense strategy.[2] It's an important assignment because over the last two administrations, we have experienced the effects of poorly conceived strategy. The result has been erosion of our strength as a nation, with stagnant incomes, declining health standards, soaring prices for the most basic ingredient of our well-being – energy – and near destruction of our financial system.[3]

---

[1] Rupert Smith, *The Utility of Force, The Art of War in the Modern World* (London: Penguin, 2005), 1. The title of this chapter is from the penultimate line of Matthew Arnold's "Dover Beach" (*New Poems*, 1867).

[2] I am using the term "national defense strategy" to mean the military component of a more comprehensive "national security strategy." The *National Security Strategy of the United States*, issued by the White House in May 2010, makes this point well: Criteria for the use of military force are considered on page 22 as one component of our larger security strategy. Find this document at http://www.whitehouse.gov/sites/default/files/rss_viewer/national_security_strategy.pdf.

[3] For a stimulating presentation, without elaborate explanation, of how to synthesize a strategy and what conceptually must be included, see Col. John Boyd's slide presentation, "The Strategic Game of ? and ?," at http://dnipogo.org/john-r-boyd/. Find there also, Boyd's other original materials. For a readable biography of Boyd's genius,

Within the Department of Defense, our strategy has eviscerated our military, burdened by a worn-out inventory of anachronistic weapons and a cadre of soldiers, sailors, marines and airmen overstressed by repeated deployments to Iraq and Afghanistan. Despite the expenditure of a trillion dollars and counting, we have failed to bring Osama bin Laden to justice or to eliminate his organization. Our efforts to install democracy in Iraq have resulted in a regime aligned with Iran and with Hezbollah in Lebanon, and our occupation of Afghanistan drags on, with no sign that we can eliminate the Taliban or reconcile Afghans to the presence of foreign infidel invaders on their soil.

Let's look at how you might make a judgment about whether the strategy document that has just appeared on your desk could make any positive change in our ability to use military forces to further the country's interests.

The World Today

A national defense strategy opens with an assessment of challenges to the United States. After you sift through the verbiage, you should be able to condense the strategist's view of the world into a few categories. If I were doing a summary of the world situation, for example, it would look something like this:

- The number of countries that possess nuclear weapons – now assumed to be nine[4] – will not decrease and may increase.
- Several states are improving their conventional (non-nuclear) military capabilities, including Russia, China and India, but these pose no threat to the United States, or to any other nuclear power. Their efforts will bolster their capabilities to deal with nearby third-rate powers and also to suppress the significant threats of internal conflict that they all face, a fact we sometimes overlook in the West:

---

see Robert Coram, *Boyd: The Fighter Pilot Who Changed the Art of War* (Little, Brown and Company, 2002).

[4] These are the United States, Russia, United Kingdom, France, China, India, Pakistan, Israel and North Korea. (*2007 Military Almanac*, Center for Defense Information, p. 26) Israel, India and Pakistan have not signed the Nuclear Non-proliferation Treaty (NPT), and North Korea has withdrawn from it.

Country	Potential internal conflicts	Miles of border
Russia[5]	Chechnya and other areas in the North Caucasus; Far Eastern border areas	12,487
China	Tibet, Taiwan, Uyghurs (potential Muslim separatists)	13,743
India	Naxalite and other Maoist guerrillas; separatist movements in Assam, Kashmir and Nagaland; sectarian violence	8,763

By comparison, the United States faces no military threat in the foreseeable future from along its 7,478 miles of border with Canada and Mexico and no internal conflict that would justify the use of military force. As you are going through the strategy, and reading the justifications of the proposed programs and their funding levels, keep asking yourself how other world powers can confront more serious threats than we face but spend significantly less money.

- All major conventional powers also possess nuclear weapons or are allies of the United States or both, and this situation will continue. Since their invention, nuclear weapons seem to have eliminated war between major powers.
- The United States could become involved in a conflict if a friendly state were attacked by another country. This is not, however, guaranteed, as Georgia learned to its detriment in 2008.
- There are any number of states that do not have functioning governments or are subject to regimes not regarded as legitimate by significant numbers of their citizens. Many of these states are also infected by insurgencies, whose goal is to overthrow such governments

[5] Russia also faces increasing internal security challenges as a result of its declining population. "Transcript of Remarks by Director of the Central Intelligence Agency, Gen. Michael V. Hayden, at the
Landon Lecture Series, Kansas State University," Central Intelligence Agency, April 30, 2008.

and replace them with themselves. Although the potential for armed conflict within and between these countries will remain high, you should ask pointed questions about why any of them poses a threat to the security of the United States.

- There are transnational non-state organizations, often called *fourth generation threats*,[6] as contrasted with sub-national insurgencies, that can do damage. Because these organizations do not possess conventional military forces of their own, they are most appropriately regarded as criminal cartels, the most immediately threatening to the United States being Mexican narco-trafficking groups and the street gangs that distribute much of their product.[7]

The first task for anyone trying to evaluate a national defense strategy is to go through it and make a set of bullet points, similar to this one. Then sit back, gather some colleagues, and look at the list. Do the potential threats and their implied priorities make sense? This is not second guessing: Much of the real activity in creating and implementing strategy takes place out of the spotlight and off the printed page. The flow of people and dollars among the elements of our defense establishment – the corporations, military services, civilian agencies and congressional committees that decide what money is spent and how – dominates national defense planning far more than any consideration of threats.

---

[6] For discussions of fourth generation warfare, see Thomas X. Hammes, *The Sling and the Stone* (St. Paul, MN: Zenith, 2004) and the various articles on the subject by one of its originators, William S. Lind. For an archive of Lind's work, see http://www.lewrockwell.com/lind/lind-arch.html. Martin van Creveld has stated that his notion of "non-trinitarian war," as described in *The Transformation of War* (New York: Free Press, 1991), is essentially the same as fourth generation war. Theorists like John Robb are examining conflict by groups so distributed, yet networked, that some have proposed a "fifth generation" of war. See Robb's book, *Brave New War* (New York: Wiley, 2007).

[7] There are some 30,000 gangs with upwards of 800,000 members in the United States. (National Youth Gang Center, "National Youth Gang Survey Analysis," 2009, accessed July 29, 2010, http://www.nationalgangcenter.gov/Survey-Analysis. Much of the income of these gangs comes from distributing the $10 to $30 billion in illegal drugs that narco-trafficking cartels move into the United States every year (Oriana Zill and Lowell Bergman, "Do the Math: Why The Illegal Drug Business is Thriving," http://www.pbs.org/wgbh/pages/frontline/shows/drugs/special/math.html). More than 26,000 people have died in drug-related violence in Mexico since December 2006, and drugs are blamed for a rising share of violence and corruption in the United States, (Randall C. Archibald, "Mexican Drug Cartel Violence Spills Over, Alarming U.S.," http://www.nytimes.com/2009/03/23/us/23border.html). The total number of al-Qaeda members in the United States is unknown, but they have been unable to mount a successful operation in this country since Sept. 11 and are not a significant threat to U.S. security at this time.

A trillion dollar annual expenditure[8] brings along this potent constituency, people whose prosperity, livelihood, or in the case of wounded veterans, for example, even survival depend on this flow of money. Moreover, in the face of gigantic deficits in 2010, the Pentagon planned for its own budget to continue to grow.[9] In addition to these players, there are any number of foreign countries and companies and their lobbying organizations within the United States who have an interest in influencing American national defense strategy.

If you decide that the strategy's picture of the world bears little resemblance to reality, then stop, document and report your findings. Clearly you are reading a political settlement among the various power centers, and you aren't interested in assessing strategies for conflict in Fantasyland. If, on the other hand, you find it reasonable – even if not what you might have written – then the next step is to consider what it is that the strategy wants military forces to accomplish. Here's an illustration of how you might proceed.

Shaping the Future

A national defense strategy does not define what the United States wants to achieve through international affairs; that is a political decision and is given to the national security apparatus by the president through a variety of channels. This is another reason why what is written in strategy papers may not be the actual strategy that the administration or the Department of Defense wishes to conduct. For example, the United States, as do all countries, has an interest in furthering the well-being of its companies abroad. Although you will find nothing in the recent "National Security Strategy" about sending in the Marines to protect U.S. commercial interests, we have done that several times, particularly in Latin America. And in the middle of its high-minded words about intervening to protect "civilians facing a grave humanitarian crisis," that same national security strategy also admits that "The United States must reserve the right to act unilaterally if necessary to defend our nation and our interests ..." whatever the administration at the time deems those interests to be.

---

[8] The total U.S. spending on all elements of national defense, including military forces, intelligence, homeland security, medical care and other services for veterans, border security, and interest on past defense-related debt will total roughly $1 trillion in 2008. See, for example, Robert Higgs, "The trillion-dollar defense budget is already here," The Independent Institute, March 15, 2007, http://www.independent.org/newsroom/article.asp?id=1941/ ; Robert Dreyfuss, "Financing the imperial armed forces: A trillion dollars and nowhere to go but up," *Tom Dispatch*, June 8, 2007, http://tomdispatch.com/post/174793/robert_dreyfuss_the_pentagon_s_blank_check/.

[9] Andrea Shalal-Esa, "Pentagon Official Sees Real Growth In Defense Budget," Reuters, May 4, 2010.

It is important, therefore, for assessing strategy to keep a grip on reality: If the desired ends are unobtainable – establish universal brotherhood and world peace, for example – it means that what you're reading is not what we intend to do. My advice, if you come to that conclusion is, again, to stop the exercise. Trying to infer our true strategy is interesting, but like the also interesting writing of alternative histories – *What if the South had held Atlanta until after the 1864 election?* – there is no way to tell if you have it right.

Philosophical Interlude: Military Force in the 21st Century

So let us assume that you find the description of both the world situation and the stated objectives at least credible. The remaining step in your assessment is to assess the following: Will the military forces that the administration proposes accomplish our objectives in the world as described? It is a question that leaders down through history have gotten wrong, so you should proceed with care and with a degree of humility.

Your judgment must rest on your conception of what military force can accomplish. This is not a purely military problem. You are not trying to predict what would happen should the United States invade "Xyz-istan" but rather to assess the usefulness of armed forces for solving problems in the modern world. You can do this, and permit me to give you some advice on how to proceed.

Although some commentators, particularly on the left, decry the creation of an American empire, the fact is that the United States has a surprisingly limited capability to influence events around the world. We can invade most any country that does not have nuclear weapons, but occupying even militarily weak countries and changing their social and political systems remains a fantasy. In 2010 we were straining, for example, to keep fewer than 200,000 troops in Iraq and Afghanistan,[10] of which perhaps 40 to 50 percent are patrolling or otherwise in combat roles (the rest perform support functions).

The cost of these operations is covered by the purchase of American debt by countries such as China, the OPEC countries and Russia in such quantities that it would be difficult to continue operations in Iraq and Afghanistan without it.[11]

---

[10] Andrew J. Bacevich, "Surging to Defeat: Petraeus's strategy only postponed the inevitable," *The American Conservative*, April 21, 2008; "No US troop increase in Afghanistan without deeper cuts in Iraq: Pentagon," Agence France-Presse, May 7, 2008.

[11] As of May 2010, China holds more than $865 billion of U.S. government securities, the "oil exporters" (i.e. the OPEC nations and others) account for another $235 billion, and Russia $126.8 billion. "Major Foreign Holders Of Treasury Securities," July 16, 2010, http://www.treas.gov/tic/mfh.txt. Some experts consider these numbers to be understated because nations can buy these securities through third-party brokers—

Yet the United States is not likely to emerge from Iraq and Afghanistan with any improvement in its national wealth. We are, it seems, the first imperial power to be paying for the privilege, with estimates of the total cost of the war running in the $3 trillion to $5 trillion range.[12] How contraction of the U.S. economy brought on by the recent recession, with the concomitant need for bailouts and stimulus packages, will affect our ability to continue paying for expensive occupations remains to be seen. Our options will be further limited by our current level of debt, which in 2011 will roughly equal our gross domestic product for the first time since the end of World War II.

Most of the means for reducing the threat from violence do not involve military forces and rely instead on trade, diplomacy, commerce, intelligence, law enforcement, tourism, educational exchange and so on.[13] In a world populated by human beings, however, there will be times when amicable agreement is not possible, when religious fervor or nationalistic feelings or a leader's ego, combined with miscalculation of the odds of success, leads to the use of force.

Granting these considerations, and others that you will think of, your task could be restated as rendering a judgment on whether the strategy defines a framework for procuring and utilizing military forces that would further our national goals as set forth in the Constitution and elaborated by the administration.

Let's look more closely at how military forces can be employed in the 21st century.

The Military Toolbox

First, there are conventional forces – the tanks, airplanes, soldiers, ships and so on like we faced in the 20th century's world wars. They can wreak enormous damage and kill huge numbers of people – fatalities in World War I numbered around 20 million, and numbers in the 50 million to 70 million range are often

---

"Caribbean banking centers," for example, hold $165.5 billion and Hong Kong holds $145.7 billion—and many regard the size and composition of their reserves as state secrets.

[12] Linda J. Bilmes and Joseph E. Stiglitz, "The Iraq War Will Cost Us $3 Trillion, And Much More," *The Washington Post*, March 9, 2008. Three trillion dollars roughly equals a $10,000 burden on every man, woman and child in the United States.

[13] The Department of Defense recognizes the "DIMES" model: diplomacy, information, military, economic and societal-cultural factors. Walter Pincus, "Irregular Warfare, Both Future and Present," *The Washington Post*, April 7, 2008. The current secretary of defense, Robert M. Gates, described this strategy in his article, "A Balanced Strategy, Reprogramming the Pentagon for a New Age," *Foreign Affairs*, January-February 2009, and the president reiterated the point in the May 2010 *National Security Strategy*.

cited for World War II – but they take some time to do it.[14] Because they need large numbers of trained troops and vast supplies of expensive weapons, they make up the majority of the world's defense budgets.

Then there are nuclear forces, which are cheap in comparison to conventional forces.[15] Like conventional weapons, nukes can cause considerable damage, but they do it in seconds. By the mid-1960s, there were enough of these in the arsenals of the major nuclear powers that the survival of the human race itself was doubtful, were they ever to be used.[16] With the total inventory now reduced to "only" many thousands, the results can unavoidably be much the same.

Finally, there is "none of the above," special forces designed to contest the "unconventional" threats that manifest themselves in "low intensity conflicts" and fourth generation (non-state) warfare. Although special forces are highly trained, there are few of them (hence "special"), and because they need little in the way of complex hardware, they are relatively cheap.

When Is the Use of Military Force Appropriate?

Because we face no direct conventional military threat to our national survival, any use of non-nuclear military forces by the United States will be voluntary,[17] the "continuation of policy" by other means, in the words of the Pentagon's favorite strategist, the early 19th century Prussian aristocrat Carl von Clausewitz.[18]

---

[14] These numbers do not include the 50 to 100 million people who died in the Spanish Flu pandemic of 1918-1920. Although the war did not cause the pandemic, conditions at the front and massive movements of troops around the world are often cited as aiding its spread and perhaps increasing its lethality. For more information, consult the Wikipedia article or http://virus.stanford.edu/uda/

[15] Just to cite one example, according to the Center for Defense Information's *2007 Military Almanac*, pages 98-99, the cost of a Trident II submarine-launched ballistic missile, capable of destroying any city on earth, is about half that of a single F-22 tactical fighter aircraft.

[16] This lesson took a while to sink in. Until about 1960, the United States considered "tactical" nuclear weapons as ordinary tools of war. See: Walter Pincus, "Eisenhower Advisers Discussed Using Nuclear Weapons in China," *The Washington Post*, April 30, 2008. In recent years the United States has had more than 5,000 nuclear warheads in its inventory.

[17] None of our mutual defense treaties require the United States to go to war if an ally is attacked. The only contractual requirement in those treaties is that we consult with the allies, not go to war. Article 5 of the North Atlantic Treaty, for example, requires a member country to "assist the Party or Parties so attacked by taking forthwith, individually and in concert with the other Parties, such action as it deems necessary," accessed July 30, 2010, http://www.nato.int/cps/en/natolive/official_texts_17120.htm.

[18] Because the German word for "policy" can also be translated as "politics," Clausewitz's formula also fits "wag the dog" wars waged for domestic political reasons.

Any national military strategy must indicate when such wars are appropriate for the United States. When, in other words, should U.S. military forces be used for missions other than the immediate defense of the United States, which requires only nuclear deterrence and very few conventional ground, sea and air forces? How the strategy answers this question determines – in theory, neglecting existing forces and spending – the size and composition of U.S. military forces. Almost without exception, anybody you discuss force structure with will have an agenda, so do your own research and think long and hard about what you find.

Can We Run on Autopilot?

Before examining potential uses for military force in the 21st century, it should be acknowledged that some people would dispense with strategy entirely, pick an arbitrary percentage of the U.S. gross domestic product, usually 4 or 5 percent, and spend that amount on something every year. The logic often provided is that we have spent that percentage and more at times in the past.[19] This rationale, however, neglects the world situations at those times, including the existence of major threats in the Soviet Union and the People's Republic of China. For your reference, at the end of the Cold War, the United States was spending 4.6 percent of GDP on defense, and it now spends 4.9 percent.[20]

Examined in this light, the arguments for holding defense spending at a constant percentage of GDP appear designed more to ensure a money flow to the defense complex than to improve the security and well-being of the rest of our citizens.[21]

Potential Uses for Non-Nuclear Military Force

Although the Soviet Union is gone, legitimate requirements for conventional and special military forces, albeit in much smaller quantities, remain.[22] You will

---

People who start wars, however, routinely experience unintended consequences, such as plagues, famines, conflicts that drag on well beyond predictions and horrendous cost overruns – not to mention losing. That the dogs of war so frequently devour those who unleash them suggests that war is anything but a rational "continuation of policy by other means."

[19] For a typical list of justifications for spending 4 percent, see The Heritage Foundation's white paper, "Providing for the Common Defense: Why 4 Percent?" The Heritage Foundation, April 2, 2008, http://www.heritage.org/Research/HomelandDefense/wp040208.cfm.

[20] A variation on this theme is to increase the defense budget each year by some percentage above inflation, again, neglecting both the external threat, or lack thereof, and the diminished utility of conventional force in a world with both nuclear weapons and increasingly sophisticated insurgencies. For a recent example, see Mackenzie Eaglen, "U.S. Defense Spending: The Mismatch between Plans and Resources," The Heritage Foundation, June 7, 2010.

[21] If you believe that more money means larger, more effective military forces, see Essay 8, "Decoding the Defense Budget," of this handbook.

find that all national defense strategies include lists of these uses. For example, in rough order of potential severity (as contrasted with likelihood):

1. A major conventional conflict – that is, one that does not go nuclear – between the United States and a "near-peer," usually hypothesized as either China or Russia.

2. Wars between states where the United States has significant interests (e.g., Saudi Arabia versus, for example, Iran or a resurgent Iraq).

3. Invasion and occupation of a developing country by U.S. military forces. Think Iraq and Afghanistan.

4. Counterinsurgency (COIN), where the military forces of the United States assist a friendly government in suppressing an attempt by indigenous rebels to overthrow it or to replace it within a region of that country, e.g. El Salvador, or – again – Afghanistan.

5. Law enforcement, where U.S. military forces assist in suppressing non-state groups other than insurgents. Anti-piracy is a topical example.

6. Stability operations, nation building and peacekeeping, where military forces are used primarily, but sometimes only initially, for non-combat roles: Somalia and Haiti.

Again, does the list offered in the strategy make sense? Are they left over from earlier strategies? Do they correlate with our current spending patterns? Are you reading another political settlement? Let's take a brief look at what your examination might include.

War Against a "Near-Peer"

As the opening quote indicates, a few strategists have come to the epiphany that the major nations, are not going to wage war on each other and so military force is of diminished utility in the modern world. The reason is not the outbreak of brotherly love but the advent of nuclear weapons.[23] Although the threat of

---

[22] Col. Douglas Macgregor, U.S. Army, ret., provides a good summary of the argument for retaining significant conventional forces in "Remember the Blitzkrieg before it's too late," *The Washington Times*, May 10, 2010 at http://www.washingtontimes.com/news/2010/may/10/remember-the-blitzkrieg-before-its-too-late/.

[23] Should some well-meaning effort succeed in eliminating nuclear weapons, conventional war between the major powers would take up where it left off.

occasional sparring cannot be ruled out, such as the Hainan P-3 incident in April 2001, you should take a hard line and ask why this most incredible scenario should dominate the sizing of U.S. conventional forces, which represent, as I've mentioned, the bulk of U.S. defense spending.

Proxy Wars and Other Wars Supporting Allies

The First Gulf War, the Korean War, and the Vietnam War were of this type: The United States itself was not threatened by foreign armies, but believed that it must intervene to help counter a third party that may itself have been supported by other major powers.

An important point about all such wars to date is that the United States did not intervene alone but formed an alliance that helped counter the attack.[24] Involvement of allies, of course, reduces the requirement for U.S. military forces, and a show of international solidarity could alleviate the need for armed intervention. You might also raise the issue of why we're always the ones trundling our military forces around the world searching for a place to replay the Battle of the Bulge. Couldn't we and our allies learn some lessons from the Vietnamese, Afghans and Iraqis that we could use in those conflicts that do pop up?

Splendid Little Wars

In the late 20th and early 21st centuries, the two outcomes that wars of choice have had in common is that they turned out to take much longer and they cost considerably more in lives and money than their proponents promised. The George W. Bush administration's estimates for the cost of the Second Gulf War, for example, were around $60 billion.[25] In the Clinton administration, the NATO-Serbian War (March 24 –June 10, 1999) was supposed to last three days, but dragged on through 78 and ended only when the alliance cobbled together the credible threat of a ground invasion and dropped demands that Serbia abdicate its sovereignty, and when the Russians withdrew their support for the Milosevic government.

Experiences such as these suggest your assessment should question any tendency to overt interventionism, at least without the support of our closest and most long-standing allies, and consider whether, if a substantial fraction of our NATO allies are not willing to join us, our proposed intervention is appropriate. Such an attitude might have served us well in the run up to the Second Gulf War.

---

[24] Even in Vietnam, we were supported by units from Korea, Australia, New Zealand and the Philippines.
[25] Widely cited on the Internet. See for example: http://www.cnn.com/2003/ALLPOLITICS/02/27/sprj.irq.war.cost/

Will COIN Theory Make Occupations Possible?

There is considerable controversy on whether counterinsurgency by outside forces – a mission sometimes known as "foreign internal defense" – is possible. The record of such attempts is not positive, El Salvador being the only recent success, and it was conducted largely through indigenous political reforms with training by U.S. forces and no direct U.S. combat involvement.[26]

The strategy you are assessing may state that counterinsurgency theory has proven itself in Iraq and so can solve the problem of other occupations. It may be early, however, to start claiming success for COIN doctrine in that country, which, two years after the "surge," continues to evolve into a religiously conservative state dominated by Shiite clerics and politicians friendly to Iran. As for the economic spoils of the war, most of these appear to be going to countries that sign the best deals with the new regime, most prominently China. Russian companies are also actively pursuing contracts in the country.[27]

Perhaps the strongest argument against strategies that posit invasion as a tool of policy, even if insurgencies against the occupation were somehow to be contained, is that nobody knows how to rebuild other peoples' destroyed societies. The area often cited as a success story, the former Yugoslavia, is an economic and social mess:

> However badly run Kosovo may be at the moment, and however much gangsterism and ethno-nationalism have flourished there under the haphazard stewardship of the so-

---

[26] U.S. assistance in El Salvador (1981 to 1992) was limited to advice, training and financial assistance. For a detailed examination of U.S. actions in El Salvador, see: Steven Metz, *Counterinsurgency Strategy and the Phoenix of American Capability* (Carlisle, PA: Strategic Studies Institute, U.S. Army War College, 1995). For an extensive discussion of counterinsurgency since the end of World War II, please refer to Chet Richards, *If We Can Keep It* (Washington: Center for Defense Information, 2008), particularly Chapter 4 at http://pogoarchives.org/labyrinth/06/01.pdf.

[27] President Bush's goals for the war included democracy and freedom for the Iraqi people, defeat of Islamic terrorists in that country, and of course, elimination of Saddam's weapons of mass destruction. Enriching Russia and China was inadvertently omitted. For a recent update on Chinese investment in Iraq's oil sector, see Leila Fadel and Ernesto Londono, "Risk-tolerant China investing heavily in Iraq as U.S. companies hold back," *The Washington Post*, http://www.washingtonpost.com/wp-dyn/content/article/2010/07/01/AR2010070103406.html. Russia's Lukoil is bidding on Iraqi contracts and state-owned giant Rosneft recently signed a joint venture with U.A.E.-based Crescent Petroleum, which has stated an interest in operations in Iraq. (Ayesha Daya and Henry Meyer, "Rosneft, Crescent Agree on U.A.E. Venture, Mull Iraq," *Business Week*, http://www.businessweek.com/news/2010-06-05/rosneft-crescent-agree-on-u-a-e-venture-mull-iraq-update3-.html).

> called international community…Bosnia is falling apart again; Macedonia still looks fragile.[28]

The upshot is that most interventions and occupations will turn out badly in the 21st century, unless brutal force to the point of depopulation is used to coerce the inhabitants into submission.[29] Again, ask hard questions about a national security strategy that implies occupying foreign lands for extended periods of time, and keep in mind that nobody you talk to knows more about how to occupy a country than you do.[30]

Law Enforcement

> *Armies will be replaced by police-like security forces on the one hand and bands of ruffians on the other, not that the difference is always clear, even today.*[31]

Much of what is hypothesized as "fourth generation warfare" – state versus non-state groups or "transnational insurgencies" – falls into this category and so does not represent a new form of warfare so much as an evolution of crime. Our opponents in these conflicts are not organized military forces or even insurgent units fighting to overthrow a government, but have more the form of transnational criminal cartels, although sometimes with an ideological or religious veneer. Like most of our probable opponents, these extra-legal organizations have neither the means nor the desire to confront our tanks and combat aircraft in conventional battle. Instead, they will move aside and blend into the population.

Once this happens, our military forces would become a minor facet of the law enforcement efforts because they cannot distinguish members of the criminal

---

[28] "The Case for Clarity," *Economist.com*, February 21, 2008.

[29] For a discussion on the need for coercion in maintaining modern occupations, see Martin van Creveld, *The Changing Face of War* (New York: Ballentine, 2006) and Sir Rupert Smith's *The Utility of Force*.

[30] You will be in good company. As this is written, DOD is also beginning to have second thoughts about the usefulness of conventional forces for counterinsurgency (all the blather about a new COIN doctrine to the contrary). See Nancy A. Youssef, "Pentagon Rethinking Value Of Major Counterinsurgencies," McClatchy Newspapers, May 12, 2010, http://www.mcclatchydc.com/2010/05/12/v-print/94058/pentagon-rethinking-value-of-major.html.

[31] Martin van Creveld, *The Transformation of War* (New York: Free Press, 1991), 225.

organization from the general population. As van Creveld also noted, the populations of developed countries do not like to see their military forces continuing to kill large numbers of villagers and wedding parties, which is the inevitable result when one cannot tell friend from foe.[32]

Stability Operations and Peacekeeping

Although the history of such operations does not give cause for optimism, the alternative – do nothing – may not be acceptable to the populations of the developed world, who sometimes demand that their military forces achieve high moral purposes, such as stopping genocide, that have nothing to do with defense of their own nations.[33] As with all incursions into the Third World, however, the stopping part may be simple compared to what follows.

What is required is establishing legitimate governments and functioning economies and their integration into the world's economic and political systems. Unfortunately, as the quote above regarding the Balkans indicates, and our experiences in Iraq, Afghanistan and Haiti reinforce, these are the very things we don't know how to do.[34]

The time-honored principle that misery loves company strongly suggests that American armed forces only participate in nonmilitary missions, including law enforcement, stability and peacekeeping, as members of coalitions.

---

[32] "To kill an opponent who is much weaker than yourself is unnecessary and therefore cruel; to let that opponent kill you is unnecessary and therefore foolish," Martin van Creveld, "Why Iraq Will End Like Vietnam Did," Project on Government Oversight, http://dnipogo.org/creveld/why_iraq_will_end_as_vietnam_did.htm. The revelation of the My Lai massacre by Seymour Hersh in November 1969 caused widespread outrage and significantly diminished support for the war. For a chronology of the massacre and subsequent events, see http://en.wikipedia.org/wiki/My_Lai_Massacre.

[33] On May 12, 2010, the U.S. Congress passed S. 1067: Lord's Resistance Army Disarmament and Northern Uganda Recovery Act of 2009 and sent it to the president for signature. As noted above, the May 2010 "National Security Strategy" explicitly endorses (on page 22) the use of military force to resolve "a grave humanitarian crisis."

[34] To illustrate, one way to jump start an economy is for the developed world to begin buying things from it. Initially, these will often be agricultural commodities. Unfortunately, such a policy runs into opposition from domestic constituencies and leads to a variety of obstacles including agricultural tariffs and subsidies, "Buy American" provisions and the desire of senior politicians to reward American contractors. For a discussion, see Thomas P. M. Barnett, *Blueprint for Action* (New York: Putnam, 2005), 244. Note that dividing a country along ethnic lines – sometimes offered as a solution for problems in developing countries – may exchange a single repressive and incompetent government for a set of them.

Conclusions

If your assessment validates the strategy, even with reservations, you're through.

If, on the other hand, you cannot avoid the conclusion that there are serious mismatches between the world situation, what we're trying to accomplish, the forces we propose to employ and the resources we intend to devote, then you have another decision to make. Do you want to report this to whoever asked you for the assessment and then go on to another assignment, or do you want to try to do something about it?

The rest of this book is intended for those taking the second alternative. It will not be the path to riches, although you may derive great satisfaction from doing good for your country. Budding national security strategists should keep a couple of points in mind before they give into despair when contemplating the enormous size, and hence momentum, of our defense-security establishment:

> First, even an aircraft carrier can be turned 180 degrees by manipulating only a tiny fraction of its structure. It has to be the right fraction.
>
> Second is the Stockdale Paradox: You must never confuse faith that you will prevail in the end - which you can never afford to lose - with the discipline to confront the most brutal facts of your current reality, whatever they might be.[35]

Although the national security establishment defeated an earlier generation of reformers,[36] you may have the great good fortune to have come on the scene at the beginning of an era of momentous change, when a trillion dollars of national security spending comes into play.[37]

---

[35] An expression coined by management guru Jim Collins in *Good to Great* (New York: Harper Collins, 2001), 83-87.

[36] For a detailed description of what happened to them, nothing beats the memoir in *Military Reform: An Uneven History and an Uncertain Future*, Winslow T. Wheeler and Lawrence J. Korb (Stanford, CA: Stanford Security Studies, 2009).

[37] Some on the political right are already talking about the need to rein in social programs, such as Medicare and Social Security, so they can preserve funding for weapon systems and standing military forces. See, for example, Mackenzie Eaglen, "U.S. Defense Spending: The Mismatch between Plans and Resources," The Heritage Foundation, June 7, 2010. On the other hand, some conservatives, such as Senator Tom Coburn, R–Okla., have proposed extracting a trillion dollars out of the defense budget by freezing it at the 2010 level. See this proposal at http://coburn.senate.gov/public/index.cfm?a=Files.Serve&File_id=3ae23727-6bbe-4ce1-8516-2b82726911cc.

Essay 7

"Follow the Money"

by Andrew Cockburn

"Follow the money," Deep Throat told Woodward and Bernstein. Endlessly and approvingly cited, these words have become a hallowed journalistic maxim, and quite right too. The problem is that most of the time this sage advice is ignored, not least by those whose job it is to report and comment on the activities of our national security system. Similarly, the venerated Dwight Eisenhower may have put the phrase "military industrial complex" in the language, but it is today deemed too loaded a term for mainstream media employment anywhere outside the opinion columns. In fact, even to suggest that U.S. military organizations exist for the benefit of those who profit from them is considered unseemly, possibly indicating a dangerous predilection for "conspiracy theories."

Instead, the public brain is more routinely softened with thoughtful ruminations such as New York Times writer Elisabeth Bumiller's July 25, 2010 article on the awesome cost of the Iraq and Afghan wars.[1] Pondering the issue, Bumiller found a partial culprit in "twenty-first century technology," as if that were a sufficient explanation and also unavoidable. It would have been helpful if the writer had looked at specific examples of the technology that is costing us so much, such as "Compass Call," a $100 million Lockheed EC-130H equipped with ground penetrating radar that searches for $25 home made bombs buried in an Afghan road – one small component of our $50 billion counter- IED (Improvised Explosive Device) effort. Readers should also be aware that those responsible for Compass Call have no excuse for believing that there is anything justifiable about it all. An in-depth study of its effectiveness in Iraq, carried by a strategic analysis "cell" of military intelligence in Baghdad in April 2007, examined the results of hundreds of flights from the previous October through to May 2007. Surveying the results, the analysts summarized them as "Conclusion: No Detectable Effect."[2]

---

[1] Elisabeth Bumiller, "The War: A Trillion Can Be Cheap," *The New York Times*, July 24, 2010, http://www.nytimes.com/2010/07/25/weekinreview/25bumiller.html.

[2] "Operational Iraq Data." Study prepared for "MultiNational Force Iraq." April, 2008. Made available to author. Estimated cost per flying hour of Compass Call is roughly $70,000.

Why We Spend

On the other hand, it is, of course, clearly a *financially* justifiable activity for the Lockheed Martin Corporation and the galaxy of subcontractors whose interests are tied to the program, a fact that should be first and foremost in the mind of anyone looking into this or any other military initiative. With “who profits?” as a *schwerpunkt* – a main objective around which all efforts are organized – analyzing the salient features of the national security state becomes a much easier and more illuminating task.

Such an approach certainly helps in understanding post-World War II U.S. history. Library shelves groan under the volumes analyzing the origins of the cold war. Recall that following victory in World War II, the U.S. rapidly disarmed, disbanding its huge conscript army and slashing weapons production. The economies of our allies and enemies in the recent conflict lay in total ruin. Although the Soviet Union controlled eastern European states overrun by the Red Army during the war, this was by prior agreement with the U.S. and Britain. Suddenly, in the spring of 1948, senior officials of the Truman Administration suddenly began issuing ominous warnings that the Soviet Union was bent on war and might attack at any time. A warning to that effect—“war could come at any time”—was solicited by the chief of army intelligence from the U.S. commander in Germany, General Lucius Clay, and duly leaked to the press.

Why?

The answer is clear for anyone who remembers to follow the money. The aircraft corporations who had garnered enormous profits during the war on the back of government contracts had discovered by 1947 that peace was ruinous. Despite initial high hopes, the commercial marketplace was proving a far harsher and less accommodating environment than that of wartime, especially as there were far more companies than required by the peacetime economy. Orders from the civilian airline industry never lived up to expectations while efforts to diversify into other products, including dishwashers and stainless steel coffins, proved disappointing and costly.

Something had to be done. In the spring of 1948 senior officials in the Truman Administration, including Secretary of Defense James Forrestal, suddenly began warning that the Soviets were on the brink of unleashing a surprise military attack against Western Europe. There was no evidence that the Soviets had any such intentions, a point, as declassified documents now make clear, that was well known to the senior officials. [3] In fact Stalin, the Soviet leader, was enjoining the powerful western European communist parties from any

---

[3] Frank Kofsky. *Harry S. Truman and the War Scare of 1948* (St Martin’s Press. 1995) 117 ff.

revolutionary action and refusing to aid the Greek communists in their civil war against the U.S.-backed government.

This cause (need for stimulus in the aerospace industry) and effect (war scare leading to sharp increase in defense appropriations) was pithily summed up at the time by Lawrence D. Bell, President of the Bell Aircraft Corporation: "As soon as there is a war scare, there is a lot of money available."[4] And so it proved. The aircraft procurement budget soared 57% as the overall Pentagon procurement budget exploded by almost 600 percent from less than $6 billion in 1947 to more than $35 billion in 1948 (in contemporary 2011 dollars). The industry, not to mention powerful institutions linked to its fortunes, such as major banks, was saved from collapse.

Apart from a brief relapse pending the outbreak of the Korean war in 1950, "war scares," otherwise known as "threat inflation" would thereafter be a regular feature of the U.S. political and economic landscape. It mattered little what the Soviet enemy was actually doing, or in a position to do. All that was required was for an announcement that "intelligence" had revealed an ominous "gap" between U.S. and Soviet capabilities, and the money flowed. The "missile gap" on which John F. Kennedy rode to victory in 1960 yielded an immediate fifteen percent hike in defense spending. Years after the money had been appropriated and spent, it was openly admitted by the relevant defense secretary, Robert McNamara, that in fact the gap had been entirely in favor of the U.S. Similar, if less infamous episodes recurred featuring bombers, tanks, ships, anti-ballistic missiles and, most comprehensively, defense budgets themselves.

Embarrassing realities, such as serious shortcomings in our putative enemies' capabilities, have generally been kept out of sight of the taxpayers. Equally, explosive cost overruns and technical disasters generate, at most, short term scandals. Pleas to cut the defense budget have rarely yielded much of a political dividend. Indeed, in former days, the very size of the budget, irrespective of its components, was touted as a necessary part of our deterrent. One of the more successful "gaps" of the cold war years was the greater size of the Soviet defense budget. The Soviets didn't announce how much they were spending on defense (even if they knew the real cost themselves, which is dubious); so the figure publicized by the military industrial complex was based on an ersatz calculation of the presumed cost to the Soviets of duplicating U.S. programs and systems. I.E., the cost of a Soviet swing-wing bomber would be assessed on the basis of the cost of a similar U.S. effort. Therefore, as Ernie Fitzgerald, the consummate Pentagon "whistleblower" of the 1960s, 1970s, and 1980s, once observed, "every time the B-1 bomber has a cost overrun, the Soviet defense budget goes up!" In other words, the more dollars we wasted, the more

---

[4] Kofsky Op cit. p170.

dangerous the other side became, which justified our wasting even more dollars, and so on.

Misguided commentators customarily referred to the cold war defense environment as the "arms race." It is important to understand that there was little or no element of military competition with the Soviets, rather one of mostly one-sided budget enhancement. This point is most easily made by comparing the level of defense spending while the U.S. was purportedly faced with a formidable and potentially aggressive enemy with the level of spending once that threat had disappeared. From 1948 to 1990, i.e. during the cold war, America spent an annual average of $440 billion (in 2011 dollars). For the period when the Pentagon budget had adjusted to the end of the cold war (that is with General Colin Powell's and Secretary of Defense Richard Cheney's "Base Force" reductions) up to the first year before the global war on terrorism (1993-2000), Pentagon spending averaged $392 billion (also in 2011 dollars). (Interestingly, during these years of the Clinton presidency, Pentagon spending was higher than the long range budget plan envisioned by Secretary of Defense Cheney.) Thus, when the Soviet Union had disintegrated and Soviet missile sub crews were offering tours of their vessels to western TV teams for $500, the US defense budget was just 11 percent lower. By subtracting the later amount from the cold wartime figure, we can discern the actual annual cost of confronting the USSR: $48 billion – tantamount to a bargain. The fact that the end of superpower confrontation made such a little difference to defense spending underlines the irrelevance of the Soviet military, save as a useful justification, to the U.S. defense system.

Clearly, military budgets are driven by something other than military requirements, at least in peacetime. But surely an actual shooting war, with American lives and vital interests at stake must be different, right? Military spending zoomed to gargantuan levels in 1950-53, for example, but those were the years of the Korean War, with almost six million men and women in uniform, of whom 140,000 were killed or wounded. That explains the huge increase in defense spending of those years? Not so. Sadly, it seems that even a shooting enemy made little difference to the way the defense system does business. Follow the money.

True, the U.S. deployed large armies to fight in the frozen rockbound wastes of the Korean peninsula – but that's not where huge amounts of the money went. The fastest growing component of the budget in those years was for "strategic" B-47 nuclear bombers (which, however, lacked intercontinental range) as well as other items useful only in a strategic nuclear war, such as the sluggish "D" version of the F-86 fighter suitable only as an anti-bomber interceptor and developing the follow on F-102 and F-106 interceptors. These, of course, were suitable only for shooting down those high altitude bombers, of which the Russians had very few, and the Koreans and Chinese none. The budget for these items soared from $2.5 billion in 1950 to $7.7 billion in 1951 to $11.3 billion in 1952.

Meanwhile, in the freezing frontline trenches, U.S. soldiers and marines lacked decent cold weather boots. Half the casualties in the first winter of the war were from frostbite. Like some threadbare guerilla army, G.I.s would raid enemy trenches to steal the warm, padded boots provided by the communist high command. "I could never figure out why I, a soldier of the richest country on earth, was having to steal boots from soldiers of the poorest country on earth," recalled one veteran of these harrowing but necessary expeditions.[5]

Lest anyone think that such outrages belong only to a dark and distant age, it is worth recalling that two years into the war in Iraq, families in the U.S. were going into debt to buy armored vests, camelbacks, socks, boots and even night vision goggles for sons, brothers and husbands whose senior commanders and congressmen and women felt no need to supply them with these items until they were embarrassed into it by the press.

In the modern era, we added $1 trillion to the defense budget after September 11, 2001 to fight the wars in Iraq and Afghanistan (up to 2010). In that same period, we added a second trillion dollars to the non-war ("base") Pentagon budget; that additional money made our Air Force and Navy smaller and our inventories of ships and combat aircraft older. In the Army, a 53 percent increase in money allowed a 5 percent increase in brigade combat teams.

How We Spend

Given this demonstration of Pentagon priorities then and now, it should come as no surprise that the torrent of money unleashed thanks to the Korean war continued to flow at only a slightly diminished rate once the guns stopped firing, with much of the money consigned to contracts for strategic systems with the "aerospace industry," as the aircraft corporations had sleekly renamed themselves. Key to the process, and to the enormous ensuing costs, was the system of "cost plus" contracts instituted in World War II that endures in one form or another to this day. So long as the contractors are guaranteed a percentage of their costs as profit, they have an obvious incentive to make those costs as great as possible. A contract to produce 100 missiles at a cost of $1 billion can yield a $50 million profit. Ergo, if it suddenly transpires that for reasons beyond man's control the cost of that program zooms to $2 billion, then the profit accordingly leapt to $100 million. It makes no difference if, as is all too likely, the cost of the individual missiles has increased so much that the $2

[5] Personal anecdote from Korean War veteran.

billion now buys only 50 missiles, or 10, or ultimately just one. The bottom line is unaffected.

In other words, as observed long ago by Ernie Fitzgerald, who battled this culture as an air force official, the contractors are "selling costs," not weapons systems. To the extent that they can improve their "products" by making them more complex and thus more expensive, they prosper. The inevitable corollary has been that the number of items produced for any one program goes down as the costs zoom up. Hence the F-35 fighter, currently under development for the Air Force, Navy and Marines as well as a number of foreign air forces, was originally slated for a production run of 2866 planes at a unit cost per plane of $81 million. Already, well before the plane has completed testing, the unit cost has soared—thus far—to $155 million each, and the total buy has accordingly shrunk to 2457. Further production cuts, as foreign buyers drop out, are inevitable, which will in turn boost the unit cost of the remaining planes on order, leading to further cuts, and so on.

Once this disconnect between the official (weapons systems of postulated quality and quantity) and actual products (costs) marketed by the defense industry is clearly grasped, other distressing aspects of the U.S. defense system become easier to understand. Escalation of costs required inefficient management practices, employing twenty people to do, supervise, manage, and administer the work of five, for example. "Inefficiency is national policy," declared the Air Force general managing the vastly over-budget F-111 bomber program in 1967.[6] But inefficient production tended to produce inefficient performance. The great missile gap fraud of the early 1960s led not only to the abandonment of all cost restraints on the crash programs instituted by the Kennedy Administration to "catch up" with the Russians, but also some egregious technical failures. The guidance system for the Minuteman II ICBM, for example, was so unreliable that 40 percent of the missiles in the silos were out of action at any one time.[7] Replacements had to be bought from the original contractor, who thereby made an extra profit thanks to having supplied faulty sets in the first place.

Since the system, despite countless reorganizations and "reforms," remains essentially unchanged in the intervening half-century, we merely have to substitute the names of today's major contracts in order to understand why our budget soars as our military shrinks, as it has. (For more details on more reform leading to more costs, see the essay entitled "Developing, Buying and Fielding Superior Weapon Systems" of this handbook; for discussion of a larger budget buying smaller (and older) forces, see "Decoding the Defense Budget.")

---

[6] A. Ernest Fitzgerald. *The High Priests of Waste* (Norton. 1972) 159. The general was "Zeke" Zoeckler.

[7] Fitzgerald, op cit, p.116.

The People Who Benefit, and Suffer

Grasping the notion that defense contractors are incentivized to maximize the costs rather than the operational capability of their products should not require much imagination. But the system requires the active complicity of soldiers, sailors and airmen who, one would think, have a direct stake in effective, reliable weapons system. The easiest way to demonstrate that the military services are nevertheless as dedicated to the maximization of costs as any corporate stockholder is to consider the fates of those who buck the system, or at least try to. Plucking just a few names from the honor roll, we can review the experience of Air Force Colonel Joe Warren, whose career was effectively ruined in the late 1960s for daring to call attention to monumental cost overruns and technical shortfalls on the C-5 program, or that of Colonel Jim Burton, forced out of the service in the 1980s for insisting that the Army redesign the Bradley Fighting Vehicle so that it would not incinerate the occupants when hit by enemy fire.[8] Even as I write, the Marine Corps is attempting to destroy the career of Franz Gayl, a former marine now a civilian working on the headquarters staff.

Gayl's offense? In 2006 he relayed pleas from the fighting troops in Iraq to Marine Corps headquarters that they be supplied with vehicles sufficiently armored to withstand the impact of increasingly lethal roadside bombs. The ubiquitous Humvee, with its vulnerable flat underbelly, offered little protection and had in fact been described as a "death trap" for this very reason in an official report following the Somali operation of the early '90s.[9] It turned out that plans to supply such vehicles, later dubbed MRAPs (Mine Resistant Ambush Protected), were already in place but were being held up because officers in the Marine Corps procurement office did not want to disrupt their arrangements with the contractor for continuing high volume Humvee production. The necessary funds had already been appropriated and no one wanted to disrupt the flow by redirecting the money to the MRAPs. Even though political pressure ultimately forced the Marines to order MRAPs, with a consequent decline in casualties, Gayl has not been forgiven, but instead subjected to further persecution by his superiors. [10]

---

[8] Read about Burton's experience in his autobiographical *The Pentagon Wars: Reformers Challenge the Old Guard* (U.S. Naval Institute Press) 1993.
[9] William C. Schneck: *After Action Report, Somalia*. Counter Mine Systems Directorate, U.S. Army Research, Fort Belvoir, VA. 1994.
[10] Statement of Franz J. Gayl, House Committee on Oversight and Government Reform, Hearings on H.R 1507, Whistleblower Protection Enhancement Act, May 14, 2009.

Clearly, impeding the progress of a procurement contract, or in any way threatening the prospects of a major program, is not the way to prosper in today's military. Taking the opposite course, on the other hand, is generally seen as key to a successful career and golden (in every sense of the word) prospects following retirement. Reviewing the career of one Air Force two star, the very model of a modern major general, enjoying a trouble-free ascent through the ranks, one caustic observer suggested the following biographic entry:

> "Under General ---'s leadership…the projected cost of the ---- program increased by several tens of billions of dollars. General --- is commended for the exemplary denial with which he approached the increasing non-executability of the program, and for the zeal with which he attacked those inside and outside the Pentagon who correctly predicted that the official schedule was hopelessly optimistic.
>
> Meanwhile, General --- further disrupted the program by focusing on the PR strategy of achieving first flight dates, regardless of whether the jets were ready for sustained testing. Under his command, the program achieved timely delivery of numerous tests assets which required major work before they were actually any use.
>
> General --- further showed his leadership qualities by bugging out, mere months before the shit hit the fan, and leaving his deputy and successor to be, inevitably, fired and publicly disgraced.
>
> Given this record, there is no reason to believe that Gen. --- will not continue to advance in rank and, on retirement, proceed to a senior post at one or other of our leading defense contractors, as so many of his fellow general officers have done before him."[11]

Once upon a time, defense contractors would reward general officers who had demonstrated their loyalty in such fashion with a well endowed corporate vice-presidency, requiring only a commitment to do their bit in lobbying colleagues still in uniform and plenty of time on the golf course. Nowadays, however, we find retirees playing more powerful corporate and influence-peddling roles – multi-tasking, as it were. The Humvee that Gayl was punished for endeavoring to supplant as the vehicle of choice in Iraq, for example, is manufactured by the AM General Corporation, headed until recently by retired four star General Paul Kern, who led the Army's Materiel Command until 2004. As well as serving as President and Chief Operating Officer at AM General, which is now controlled by billionaire Ron Perelman's MacAndrew & Forbes holding company, Kern

---

[11] The author of this "bio" has spent decades as an intimate observer of the Pentagon; the deletion of the Major General's name is not to protect the guilty protagonist, but the innocent source.

was also welcomed onto the board of the EDO Corporation, a lead contractor in the burgeoning counter-IED electronics industry. When EDO was bought by ITT, Kern transitioned to the merged corporate board, having also served on the board of IRobot, manufacturer of some ubiquitous counter-IED robots, as well as CoVant Technologies, a private equity group specializing in defense investments in the Washington area.

Kern's involvement with firms associated with the counter-IED mission serves as a reminder that whereas once upon a time the military industrial complex depended on "scares" generated as needed by our impressively large Soviet adversary, today's conflicts with lightly armed insurgents offer rewards that are hardly less fulfilling. "Asymmetric warfare" has turned out to be even more expensive and at least as rewarding. Not only has annual Pentagon spending gone up tremendously above cold war levels since September 11, 2001, but also the lowly home-made bomb, or IED, occupies a place in the threat pantheon once reserved for the likes of Soviet ICBMs. Thus far, the Pentagon's Joint IED Defeat Organization has spent at least $50 billion in countering these garage-made threats, and, despite increasing US deaths from IEDs and a rising chorus of criticism, there is no sign the spigot is being turned down in any meaningful way.

The rise of CACI, a northern Virginia corporation serves as an instructive case study of the beneficiaries of today's threat environment, in which a corporation can rise to great prosperity (with a headquarters building emblazoned with its titular acronym looming over I-66 on the approaches to Washington D.C.) without actually making anything at all. Its functions, as a close scrutiny of the CACI website reveals, being in the unexplained area of "analysis" and "support"—a pure example of "selling costs." Originally intended by its founders to commercialize their SIMSCRIPT simulation programming language, the war on terror brought many fresh opportunities to CACI, including a contract to supply interrogators for the notorious Abu Ghraib jail. Though that service does not today appear in the list of employment opportunities on offer on the company's website, there are no lack of listings for work subcontracted by the Joint IED Defeat Organization, which remains much beloved by the service bureaucracies and their corporate partners for its mandate to apportion funds without specific authorization.

The CACI website also helpfully lists the board of directors, complete with biographies, thereby furnishing a useful cameo of today's military industrial complex. Topping the list of outside directors is Gordon England, best known for his service as Navy Secretary and Deputy Secretary of Defense in the George W. Bush Administration, in which capacity he adroitly avoided the odium incurred by Donald Rumsfeld and displayed a helpful solicitude for the interests of major contractors, ever ready to run interference with Congress on their behalf.[12] That was hardly surprising, given England's prior service with the General Dynamics, Lockheed, Litton and Honeywell Corporations.

Another name that catches the eye is the retired and superbly well connected four star Admiral Gregory Johnson, who earned the trust of his peers not only as the commander of far-flung fleets, but also as senior military assistant to Secretary of Defense William Cohen. Meanwhile, James L. Pavitt, formerly Deputy Director for Operations of the CIA, where, the biography informs us, he led the agency's "operational response" to the 9/11 attacks, clearly makes a good fit on the board, as does retired four star army general William Wallace, who commanded a corps during the 2003 invasion of Iraq before ascending to the command of the army's Combined Arms Center and ultimately the potent Training and Doctrine Command. Interestingly, Wallace's CACI biography cites his role in developing the Future Combat Systems, a $160 billion baroque extravaganza infamous for monumental overruns and technical catastrophe and ultimately cancelled, but perhaps in such circles this is seen as a recommendation.

Also on the CACI board sits James Gilmore, former governor of Virginia, whose biography is larded with references to his experience in the bountiful area of homeland security. Dr. Warren Phillips, a former academic with a expertise in oil pipelines and armored vehicles, along with a lawyer and a graduate of the railroad and natural gas industry round off the roster of this truly twenty-first century defense company, with 2010 sales in excess of $3 billion.

No survey of the relationship between the corporate and military professions would be complete without comparing the differing fates of General John M. Keane and Admiral William J. Fallon. Both rose to dizzying heights in the military command structure; Keane retired as Vice Chief of the Army while Fallon was head of Central Command. In his latter years in the service, Keane shared the doubts of his fellow generals regarding the Iraq adventure, but kept his thoughts to himself, maintaining good relationships not only with Defense Secretary Donald Rumsfeld and Vice President Richard Cheney, but also with other politically significant factions in the corporate, political, and media worlds. Keane has long accepted a major share of the credit for conceiving the notion of a "surge" in Iraq – now deemed the key to victory – though the all-important concept of buying off the insurgents would seem to have originated elsewhere. Keane has since become a highly sought after talker, advisor and

---

[12] One example suffices: A well informed critic of the lethal V-22 boondoggle was giving a scheduled briefing to an influential congressman on the drawbacks to the program, notably its tendency to kill the marines who were riding in it. Who should drop in, "just passing by," but Mr. England, who enquired on the topic of discussion and then weighed in with what was obviously a very carefully prepared rebuttal, defending the V-22.

policy guru. Today he also sits on the board of General Dynamics (to which he made a swift ascension after retiring) and many other boards, including Ron Perelman's MacAndrews and Forbes, is a senior adviser to the private equity giant Kohlberg, Kravis, & Roberts, sits on the board of the Rand Corporation, comments on security matters for ABC News, and is generally a potent force in today's military industrial complex.

Admiral Fallon, on the other hand, today sports only a few comparatively insignificant corporate appointments on his CV. The point of departure in the career trajectories of the two men would appear to have been Admiral Fallon's public and private outspokenness on a variety of subjects, including his rejection of the notion that Iran posed a significant threat to the U.S., coupled with spirited denunciation of a pre-emptive U.S. attack on Iran when that thinking was de rigueur in the George W. Bush administration. Such defiance of the Washington consensus, especially in an area where precise correctness is required among neo-cons and other supporters of Israel, got Fallon promptly fired and dispatched to the wilderness by George W. Bush.

A review of a hundred leading defense corporate boards would uncover many similar instances of the close embrace between the senior officer class (along with their intelligence colleagues) and the industries that serve them. That is one more reason why, in considering policies and priorities of our military leadership, outside observers must never lose sight of the pond in which they swim.

Whether it be the enduring phenomenon of the neo-cons, a group originally fostered in the mid 1970s by the late Paul Nitze as a means to enlist Israel supporters in the cause of bigger defense budgets, or the specter of the (alleged) Iranian nuclear weapons program that has so far generated $123 billion worth of U.S. weapons sales in the region, or any other aspect or issue related to U.S. national security, Deep Throat's sage advice should always be in the forefront of a truly enquiring mind.

Essay 8

"Decoding the Defense Budget"

by Winslow T. Wheeler

Many in Congress and journalism hold some commonly accepted assumptions about defense spending. Among them are that the Department of Defense budget represents U.S. national security spending, that senior DOD officials can be relied on to fairly and honestly interpret the Pentagon budget, that Pentagon numbers for the cost of programs and policies are valid and authoritative, and that more money means more defense. Perhaps, these presumptions are better characterized as hypotheses to be tested. Doing so can yield a fuller understanding of the defense budget.

The first question is –

What Is the Defense Budget?

Each year in early February, the Pentagon releases what is invariably called the "defense budget" in press articles. The numbers presented do not address all forms of defense spending; they do not even address all forms of Pentagon spending.

For example, a table included in the Pentagon's press materials for the 2011 budget shows the "base" (non-Iraq or -Afghanistan war) budget request at $549.8 billion.[1] The materials presented by the Office of Management and Budget (OMB) are more complete. The 2011 budget request for "base" (non-war) Pentagon spending was $554.1 billion. The additional $4.3 billion was for "mandatory" spending (also known as "entitlement" spending) mostly for personnel programs. The number the Pentagon released was for the "discretionary" (new annual appropriations) spending. The difference may be a minor one in this case, but it can be significant; in past years Congress has added scores of billions in new mandatory spending for military healthcare, and retirement and survivors' benefits.

---

[1] See page 8 of the material presented to the press on Feb. 1, 2010 purporting to describe the request for the 2011 DOD budget at http://www.defense.gov/news/FINAL%20PRESS%20RELEASE%20v3%20%201.pdf.

The more complete exposition of DOD budgets in the OMB materials is not easy to find; it is usually buried in the "Supplemental Materials" to a volume called *"Analytical Perspectives"* that is released each year the same day the Pentagon releases its version of its budget. Unfortunately, the DOD press corps roundly ignores the more complete OMB materials. To be better informed in future years, track it down.[2]

The same OMB table yields other important information: the additional DOD spending requested for the wars in Iraq and Afghanistan, not just for the budget year but also for succeeding "out-years," and the non-DOD spending for what OMB calls the "National Defense Budget Function," which includes nuclear weapons, the Selective Service, the National Defense Stockpile of minerals and commodities, and more. The total for 2011 comes to $738.7 billion in "total" (discretionary plus mandatory) spending.

The same table also yields the budget amounts for the departments of Homeland (domestic) Security, State (for economic and weapons aid and other national security programs) and Veterans Affairs (for what might be called the human cost of wars). Each is clearly related to national security or "defense," writ broadly. Finally, if you know where to look near the bottom of this long OMB table, you can find some additional spending in the Treasury Department for military retirement and healthcare, and finally the data needed to make a calculation of how much of the 2011 payment for interest on the national debt can fairly be attributed to the Pentagon.

The results of this more complete compilation of the president's 2011 budget request for "defense" is summarized in Table 1 below.

---

[2] Find the 2011 version of this OMB table at http://www.whitehouse.gov/omb/budget/fy2011/assets/32_1.pdf .

Table 1: Defense Related Budget Requests for 2011.

Spending Category	**President's 2011 Budget Request (in $ billions)**
"Base" DOD Budget (Discretionary only)	548.9
DOD (Mandatory only)	4.3
DOD War Spending	159.1
DOD Total	**712.3**
DOE (Defense)	18. 8
Miscellaneous Defense -Related Agencies	7.6
National Defense Budget Function Total	**738.7**
Homeland Security (DHS)	43.6
Veterans Affairs (DVA)	122.0
International Affairs	65.3
Treasury Dept. Military Retirement Payments	25.9
Interest on DOD Retiree Health Care Fund	5.7
19% of Interest on Debt (DOD Proportional Share)	47.7
Grand Total	**1,048.9**

The next time someone tries to tell you that the numbers DOD throws at you in its press releases are what you should use to understand monies spent for national security, give him a polite smile; then, go to that obscure table in the Supplementary Materials in OMB's "*Analytical Perspectives.*" It is published online the same day as the Pentagon press release. A few minutes of checking can give you a more complete understanding than what the press will report.

Selling the Pentagon Budget

Once the numbers are distributed by Pentagon press releases and their spawn in most news articles, large amounts of energy are spent in Washington to shape how they are understood. Those wanting to increase the DOD budget try to make it seem smaller by playing a popular game—the "Percent of Gross Domestic Product" diversion.

In 2007, for example, some Pentagon leaders, including the Chairman of the Joint Chiefs of Staff, told the press that the nation needed to increase Pentagon spending from 3.3 percent of the U.S. Gross Domestic Product to 4.0 percent. This argument was also frequently heard from DOD budget growth think tanks, like The Heritage Foundation. An increase of 0.7 percent should not be much, especially if, as they said, we spent much more during the Cold War—such as the 8.9 percent we spent in 1968 during the Lyndon Johnson administration.

In 2007, the Gross Domestic Product was $13.4 trillion. If we increased the Pentagon's "share" of it from 3.3 to 4.0 percent, that 0.7 percent increase meant $94 billion more for the Pentagon. What sounded like a tiny increase turns out to be significant.

The GDP has been going up tremendously over time and even though the percent of GDP that we spend on defense has gone down, the actual dollar amount we spend on defense has been going up. In fact, today we spend more on defense than we did at any time since the end of World War II; that's in inflation adjusted dollars, and it's according to the Pentagon's own official budget data in something called "National Defense Budget Estimates," more popularly known as the "Green Book."

The Green Book is an essential tool for anyone working with Pentagon spending numbers. It is annually updated and presented at the DOD Comptroller's Web site.[3]
Figure 1 below shows the DOD budget history (from the Green Book); it is a lot more informative than politically driven assertions with the effect of making the biggest ever appear to be the smallest ever.

---

[3] Find the 2011 edition of the Green Book at http://comptroller.defense.gov/defbudget/fy2011/FY11_Green_Book.pdf. If you check the data out for 2011, don't be deceived; the amount does not include all costs for the wars in Iraq and Afghanistan; although such costs are included – for the most part – for previous years. For unexplained reasons, the DOD Comptroller's Office chose to cite an incomplete figure for 2011. There's more on the frailty of this much used, important document below.

Figure 1: DOD Budget Authority 1948-2011, in $ Billions of Constant 2011 Dollars

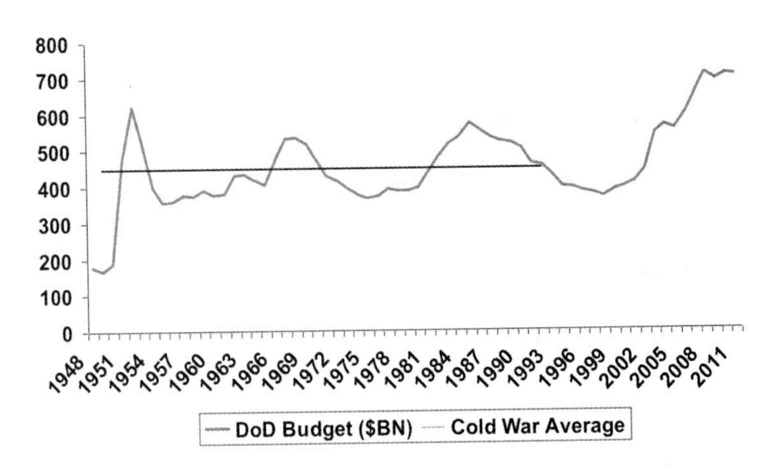

According to the Green Book, the previous high point in post-World War II Pentagon spending was 1952 – during the Korean War – at $622.9 billion (see page 109). President George W. Bush's Pentagon budgets were higher (peaking at $708.5 billion (see page 114), and are now being matched, and outdone, by President Obama.

If you understand the GDP game, you'll understand that you can make the Pentagon budget appear to grow while it is actually shrinking: in a recession, the economy might shrink by two percent; if the Pentagon budget shrinks by the lesser percentage of one percent, it will appear to be growing according to this misleading measure.

What Do Weapons Cost?

On Wednesday March 25, 2009, an F-22 crashed near Edwards Air Force Base in California. Sadly, the pilot was killed. The news articles surrounding this event contained some strange assertions about the cost of the crashed airplane.

Based on the price asserted in the Air Force's "fact" sheet on the F-22 that was linked to a Pentagon news release on the crash, the press articles on the crash cited the cost per aircraft at $143 million.[4]

It was incomplete, to put it charitably, but the media passed it on nevertheless.

The extant "Selected Acquisition Report" (SAR) from the Defense Department is the definitive DOD data available to the public on the costs for the F-22.[5] The SAR showed a "Current Estimate" for the F-22 program in "Then-Year" dollars of $64.540 billion. That $64.5 billion was for 184 aircraft.

Do the arithmetic: $64.540/184 = $350.1. Total program unit price for one F-22 calculates to $350 million per copy.

So, where does the $143 million unit cost come from? Many will recognize that as the "flyaway" cost: the amount we pay today, just for the ongoing production costs of an F-22. (Note, however, the "flyaway" cost does not include the pilot, fuel and other consumables needed to fly the aircraft away.) The SAR cost includes not just procurement costs, but research and development (R&D) and some military construction, as well.

At about the same time as the crash, a massive lobbying effort had started to buy more F-22s, to reverse Secretary of Defense Robert Gates impending announcement (in April 2009) that he wanted no more. F-22 advocates were asserting the aircraft could be had for this bargain $143 million unit price. That was, they argued, the "cost to go" for buying new models, which would not include the R&D and other initially high production costs already sunk into the program.

Congressional appropriations bills and their accompanying reports are not user-friendly documents, but having plowed through them for decades, I know many of the places and methods that Appropriations Committee staff like to use to hide and obscure what Congress and the Pentagon are actually spending. Let's check through the 2009 congressional appropriations for the F-22. Most – but not all – of the required information is contained in HR 2638, which contained the Department of Defense Appropriations Act for fiscal year 2009.[6]

---

[4] Find the U.S. Air Force "fact" sheet at http://www.af.mil/information/factsheets/factsheet.asp?id=199.

[5] Find the SAR at http://www.acq.osd.mil/ara/am/sar/2008-SEP-SARSUMTAB.pdf.

[6] Find the Act and related documentation at the definitive public Web site for bills and laws, the Library of Congress' "Thomas" Web site, http://thomas.loc.gov

In the "Joint Explanatory Statement" accompanying the bill, the House and Senate appropriators specified that $2.907 billion was to be appropriated for 20 F-22s in 2009. The math comes to just about what the Air Force said, $145 million per copy. So, what's the problem?

Flipping down to the section on "modification of aircraft" we find another $327 million for the F-22 program.

Switching over to the Research and Development section, we find another $607 million for the F-22 under the title "Operational System Development."

Some will know it is typical for DOD to provide "advance procurement" money in previous appropriations bills to support the subsequent year's purchase. In the case of the 2009 buy of 20 F-22's, the previous 2008 appropriations act provided "advance procurement" for "long lead" F-22 items to enable the 2009 buy. The amount was $427 million.

Here's the math: $2.907 + $.327 + $.607 + $.427 = $4.268 billion for 20 aircraft. That's $213 million each.

Do not think these data represent an exceptional year. If you check any of the annual buys of F-22s, you will find the same pattern: in addition to the annual "procurement" amount, there is additional "modification," R&D" and advance procurement.

A few weeks later, F-22 advocate Sen. Saxby Chambliss, R–Ga., attempted to amend the 2010 DOD "authorization" bill coming out of the Senate Armed Services Committee to buy seven more F-22s for $1.75 billion, or $250 million each. The Chambliss effort, almost certainly worked out in close association with Lockheed Martin – a major F-22 plant is in Marietta, Ga. – surely sought to pay Lockheed the full amount to procure more aircraft: not $143 million each, but $250 million.[7]

Clearly, Chambliss and Lockheed knew about some additional F-22 costs not included in my estimate of $213 million.

The pathology of low-balling a weapon's costs goes far beyond the F-22 example cited here; it is a basic tenet of bureaucratic behavior; it helps a program acquire support by top DOD management and Congress.

---

[7] Find data, discussion and sources about this failed effort in an essay "How Congress Finances Pork" at http://www.cdi.org/friendlyversion/printversion.cfm?documentID=4535.

Understatement of cost does not occur in isolation in the Pentagon; it is accompanied by an overstatement of the performance the program will bring, and the schedule articulated will be unrealistically optimistic. Once the hook is set in the form of an approved program in the Pentagon (based on optimistic numbers) and an annual funding stream for it from Congress (based on local jobs and campaign contributions), the reality of actual cost, schedule and performance will come too late to generate anything but a few pesky newspaper articles. It is a system described in great detail in a 1990 essay, "Defense Power Games" by Franklin C. Spinney. He elaborated on the same themes in subsequent testimony to Congress in 2002.[8]

Does More Money Mean More Defense?

For as long as I can remember, the politicians in Congress, the Pentagon and think tanks have judged whether people are "strong" or "weak" on defense using money as the measure. If you want to increase the budget, you are "pro-defense"; if you want to take money away, you are "anti-defense." It is that simple.

Having attacked Democrats for decades on precisely this basis, the Republicans have trained them to be shy about Pentagon budget cuts. The Democrats, especially self-proclaimed moderates, have favored Pentagon budget increases to protect themselves, but – alas – the Republican attacks don't stop; instead, they assert the Democrats' increases are too tepid. Moreover, contractors build support networks for their weapons programs by spreading dollars, jobs and profits to as many congressional districts as possible. Thoroughly imbued by their superficial view of military strength combined with the money flowing to their states and elections campaigns, senators and representatives of both political parties have failed to notice that more spending merely set the stage for a meltdown.

With the Pentagon at post-World War II spending highs, the U.S. Navy has fewer combat ships than in any year since 1946; the U.S. Air Force has fewer combat aircraft, and the Army hit a post-World War II low in combat division-equivalents in 2008 – from which it has barely recovered. These trends are not new; as the defense budget has grown over time, our forces have shrunk.

The current DOD plan is to make the Air Force even smaller, at dramatically increased cost. The Navy has a 2010 shipbuilding plan to marginally increase the fleet from 287 ships to just over 300, if shipbuilding budgets are increased by 30 percent or more. Meanwhile current plans to "upgrade" naval aviation

---

[8] Find both of these analyses at http://pogoarchives.org/labyrinth/01/09.pdf and http://pogoarchives.org/labyrinth/01/02.pdf.

mean a smaller force than we had during the Clinton administration but at a cost of more than quadrupling spending for naval combat aircraft. The Army has been spending up to $88 billion to increase its active-duty combat brigades from 38 to just 42.[9]

As they have shrunk, major hardware inventories have simultaneously grown older, on average. According to the Congressional Budget Office (CBO), among others, major categories of military equipment are aging at unprecedented rates. CBO data also shows us that the DOD plan in many major hardware categories is to make this problem worse.[10]

Data from the Pentagon show that significant elements of our armed forces are far less ready for combat than they should be. Air Force and Navy combat pilots get one-half to one-third of the in-air training time they had, for example, in the Vietnam War era. Army units are sent into Iraq and Afghanistan without the months of training, and re-training, they need, and what training they get almost never includes all the equipment and people they will take with them into combat.[11]

Some argue that, while these trends should be reversed, they are not a cause of major alarm because American "high tech" compensates for the smaller inventory. As was the case in Vietnam, the immeasurable technological advantage we hold over our enemies in Iraq and Afghanistan means little to nothing in winning the form of conflict we find ourselves in. In fact, one of our most advanced, "high-tech" systems in Afghanistan and Pakistan – the unmanned drones used to attack the al-Qaeda and Taliban leadership – is very clearly a double-edged sword. The successful attacks against more than 600 militants (according to one tabulation) notwithstanding, the drones may be helping the enemy more than us. In Pakistan, drones are said to have also killed over 300 noncombatant civilians.[12] In Afghanistan, the available data is

---

[9] For more discussion of these data and trends see Chapter 11, "Understand, Then Contain America's Out of Control Defense Budget," *America's Defense Meltdown* (Stanford University Press, 2009),

[10] For example, see "The Long-Term Implications of Current Defense Plans: Detailed Update for Fiscal Year 2008," Congressional Budget Office, March 2008, http://cbo.gov/ftpdocs/90xx/doc9043/03-28-CurrentDefensePlans.pdf.

[11] While data on training tempos are typically classified, some data for U.S. Air Force F-16 and U.S. Navy F-18 flying hours can be found from their budget presentations, such as for 2010 at http://www.saffm.hq.af.mil/budget/ and http://www.finance.hq.navy.mil/fmb/10pres/books.htm. Data for F-22 flying hours was collected during visits by the author to Langley and Nellis Air Force bases in 2006. Data for Vietnam-era flying hours is based on information collected at the time from pilots and DOD officials.

[12] See Peter Bergen and Katherine Tiedmann, "The Year of the Drone; Analysis of US Drone Strikes in Pakistan, 2004-2010," The New America Foundation,

confounded by civilian deaths also incurred by U.S. and NATO manned aircraft, but the numbers are unavoidably troubling. In a region where the code of honor demands avenging the deaths of family members and foreigners are already reviled, each drone-caused civilian death is likely to inflame multiple reactions against us.

For waging conventional war, the new weapons we buy to replace existing ones increase in cost far faster than the budget increases (which makes inevitable the shrinking and aging of our weapons, at growing cost).

Also, the new systems rarely, if ever, bring a performance improvement commensurate with the cost increase. In some cases the new system is even a step backwards. The F-35 Joint Strike Fighter is a good example. Among the aircraft it is to replace is the 1970s vintage – but still much used and almost universally praised – A-10 close air support aircraft. Even if the F-35 stays at its 2010 purchase price of over $150 million per aircraft (which it will not), it will cost ten times more than an A-10. For that additional expense, it will have less payload than an A-10; it will not be able to loiter over the battlefield to help troops engaged in combat hour after hour; it will be too fast to be able to find targets independently, and it will be too fragile and sluggish to survive at the low altitude it must operate at to be effective, even against the primitive small arms and machine gun defenses terrorists and insurgents can mount. To make matters worse, the F-35 will lack the extraordinarily effective 30 mm cannon the A-10 carries.[13]

The conventional wisdom that more money means more defense is superficial political hype.

---

http://www.newamerica.net/publications/policy/the_year_of_the_drone. See also "CIA Drone Operators Oppose Strikes as Helping al-Qaeda," Gareth Porter, AntiWar.com, June 4, 2010.

[13] For further discussion, see "Joint Strike Fighter: Latest Hot Spot in Americas Defense Meltdown," *Jane's Defence Weekly*, September 8, 2008, at http://www.cdi.org/program/document.cfm?documentid=4370&programID=37. For cost comparison data see "How Much Will Each F-35 Cost?" at http://www.cdi.org/friendlyversion/printversion.cfm?documentID=4596&from_page=../program/document.cfm and "Still More F-35 Cost Growth to Come," at http://www.cdi.org/program/document.cfm?DocumentID=4599&StartRow=1&ListRows=10&appendURL=&Orderby=D.DateLastUpdated&ProgramID=37&from_page=index.cfm. For A-10 cost data, see p. 166 of "Operation Desert Storm: Evaluation of the Air Campaign," U.S. General Accounting Office, June 1997, GAO/NSIAD-97-135, http://www.gao.gov/archive/1997/ns97134.pdf.

Other Problems in DOD Data

The DOD "Green Book" ("National Defense Budget Estimates for FY 20XX") is extremely useful, albeit flawed.[14] It contains tables on discretionary Pentagon appropriations sliced and diced several different ways. It also contains DOD manpower (civilian and military) figures, some basic economic data (such as work force, and budget ratios) and the multipliers to convert older dollars to current dollars. It is released annually by the DOD Comptroller.

The numbers presented for contemporaneous budget years need to be treated with caution, however. The budget year data does not typically include spending for ongoing wars, but the previous years will—mostly. Sometimes, the year just before the budget year will also be incomplete because a new supplemental has been requested but has not yet been acted on by Congress.

There are also some systemic problems in the Green Book. In the 1980s, Chuck Spinney and Pierre Sprey uncovered an ongoing enterprise in DOD regarding inflation that had the effect, and very probably the intent, to mask cost growth in DOD programs as inflation and to hide enormous budget windfalls (in excess of $30 billion) in appropriations that assumed excessively high predictions of future inflation. DOD applied cooked inflation indices to procurement spending and incorporated its dubious economic assumptions into the data in the Green Book. As Spinney found, the constant dollar calculations used in DOD's inflation indices can exaggerate the effects of past inflation. This can elevate past years relative to current ones and make today's spending appear less of an increase than would otherwise appear. Even though GAO investigated the matter in the 1980s and found DOD's inflation indices to be flawed, the biased calculations were never backed out of the Green Book. Doing so is surely possible, but it would also require a direct intervention by the DOD Comptroller, which is unlikely under current circumstances. While DOD's spending estimates in the Green Book are the most authoritative available, keep in mind that they are also flawed.

DOD's SARs (Selected Acquisition Reports) are also an important, but flawed, tool.[15]

The summary tables list DOD's current estimates for the cost to acquire a major defense acquisition program, including the procurement, development (including research, testing and evaluation) and military construction costs. There are several limitations, however:

---

[14] Find it at the Web site of the Office of the Under Secretary of Defense (Comptroller) at http://comptroller.defense.gov/.
[15] Summary tables of SARs are available at http://www.acq.osd.mil/ara/am/sar/.

- Many important programs, such as infantry rifles, are not included; they do not trip the spending criteria to qualify as a "major defense acquisition program."

- Sometimes there are costs associated with the program that DOD has arbitrarily excluded from the SAR estimate.

- Frequently, DOD will re-adjust the "base year" of the program which has the effect of removing the appearance of cost growth. For example, the December 2009 SAR shows the base year of the Air Force's F-22 and the Marines' V-22 both to be 2005. In fact, both programs started in the 1980s and showing the initial cost estimates exposes the huge cost growth both programs have experienced.

- No support costs are shown. Inherent in the cost of any program, expenses to maintain, operate and provide spare parts and training for any program are essential to understand the cost to possess the equipment. Such data is usually hard to find in the unclassified world.[16]

Finally, and most importantly, no one checks the SARs seriously. GAO does not automatically review them and recommend, or perform, recalculations or revisions. More importantly, the programs the SARs assess are not audited. The spending figures are not verified; they are merely contemporary estimates.

The Ultimate in Cooked Numbers

Late in 2009, the DOD Inspector General (DOD IG) reported the following in its "Summary of DOD Office of the Inspector General Audits of Financial Management."[17]

---

[16] Unclassified data available to the author for the cost to operate and support aircraft vividly demonstrates that modern, more complex, equipment is invariably much more expensive to operate than the equipment it replaces, even when that older equipment is quite ancient and needs constant work. For example, while Air Force data shows 1950s vintage B-52 H bombers to cost a hefty $2.7 million each to support annually; a much newer B-2 costs $9.7 million on the same measure. It is also true that typical DOD support costs do not include all support costs. Never included is the government overhead, such as operating the System Program Office; paying for contract bid preparation, response, and evaluation; DOD analysis of systems at CAPE and other offices, and more. One experienced senior contractor employee "guesstimated" that such costs would add as much as 3 percent to total program cost.

[17] Find this report at the DOD IG Web site at http://www.dodig.mil/audit/reports/fy10/10-002.pdf.

- The financial management systems DOD has put in place to control and monitor the money flow don't facilitate but actually "prevent DOD from collecting and reporting financial information … that is accurate, reliable, and timely." (p. 4)

- DOD frequently enters "unsupported" (i.e. imaginary) amounts in its books (p. 13) and uses those figures to make the books balance. (p. 14) Inventory records are not reviewed and adjusted; unreliable and inaccurate data are used to report inventories, and purchases are made based on those distorted inventory reports. (p. 7)

- DOD managers do not know how much money is in their accounts at the Treasury, or when they spend more than Congress appropriates to them. (p. 5)[18] Nor does DOD "record, report, collect, and reconcile" funds received from other agencies or the public (p. 6), and DOD tracks neither buyer nor seller amounts when conducting transactions with other agencies. (p. 12)

- "The cost and depreciation of the DOD general property, plant, and equipment are not reliably reported …." (p. 8); "…the value of DOD property and material in the possession of contractors is not reliably reported." (p. 9)

- DOD does not know who owes it money, nor how much. (p. 10.)

It gets worse; overall –

- "audit trails" are not kept "in sufficient detail," which means no one can track the money;

- DOD's "Internal Controls," intended to track the money, are inoperative. Thus, DOD cost reports and financial statements are inaccurate, and the size, even the direction (in plus or minus values), of the errors cannot be identified, and

- DOD does not observe many of the laws that govern all this.

That last finding is perhaps the most appalling. It is as if the accountability and appropriations clauses of the U.S. Constitution were just window dressing,

---

[18] Technically, this is a violation of the Anti-Deficiency Act, a statute carrying felony sanctions of fines and imprisonment.

behind which this mind-numbing malfeasance thrives. Congress and the Pentagon annually report and hold hearings on DOD's lack of financial accountability and sometimes enact new laws, but many of the new laws simply permit the Pentagon to ignore the previous ones; others are eyewash.

For example, the DOD IG reports that "The Chief Financial Officers Act of 1990 ... required ... [DOD] to prepare ... financial statements that were audited by either the Inspector General or an independent public accountant .... Beginning in 1991, DOD began preparing and submitting financial statements for audit. However, DOD OIG audits of those financial statements for FYs 1991 through 2001 identified pervasive and long-standing material weaknesses which caused those financial statements to be un-auditable. As a result, Congress passed the 'National Defense Authorization Act for Fiscal Year 2002,' on Dec. 28, 2001, that limits the amount of audit work performed by the DOD OIG under the CFO Act based on management's representation regarding the [un-]reliability of the financial statements." (See p. 1 of the summary report identified above.)

In other words, the 1990 law aimed at imposing accountability was waived by the update, permitting the Pentagon to ignore its statutory and constitutional requirements. This was done by both Democratic and Republican administrations and Congresses. The behavior continues to this day. The recently enacted National Defense Authorization Act for 2010 contains a Section 1003 ("Audit Readiness of Financial Statements of the Department of Defense") which instructs DOD management to produce a plan "ensuring the financial statements of the Department of Defense are validated as ready for audit..."[19] The plan is not to be effected until Sept. 30, 2017. The DOD Comptroller, the Department's CFO, has let it be known that he will seek relief even from this extremely relaxed deadline.

If you have a system that does not accurately know what its spending history is, and does not know what it is now (and does not care to redress the matter), how can you expect it to make a competent, honest estimate of future costs? It is self-evident that an operation that tolerates inaccurate, unverifiable data cannot be soundly managed; it exempts itself from any reasonable standard of efficiency.

Recall, also that the errors in cost, schedule and performance that result are not random: actual costs always turn out to be much higher than, sometimes even multiples of, early estimates; the schedule is always optimistic, and the performance is always inflated.

---

[19] See p. 801 of the Conference Report for HR 2647, National Defense Authorization Act for Fiscal Year 2010, http://frwebgate.access.gpo.gov/cgi-bin/getdoc.cgi?dbname=111_cong_reports&docid=f:hr288.111.pdf.

The Pentagon, defense industry and their congressional operatives want—need—to increase the money flow into the system to pretend to improve it. Supported by a psychology of excessive secrecy, generated fear and the ideological belief that there is no alternative to high cost, high complexity weapons, higher budgets are easier to justify, especially if no one can sort out how the Pentagon actually spends its money.

The key to the DOD spending problem is to initiate financial accountability. No failed system can be understood or fixed if it cannot be accurately measured.

And yet, there is no sense of urgency in the Pentagon to do anything about it. Indeed, in the 1990s, we were promised the accountability problem would be solved by 1997. In the early 2000s, we were promised it would be solved by 2007; then by 2016; then by 2017. Now we are being told that to expect a fix in 2017 is unrealistic.

Conclusion

The question must be asked: if nothing has been done by the Pentagon to end the accountability problem after more than 20 years of promises, is top management simply incompetent, or is this the intended result of obfuscation to avert accountability?

A spending system that effectively audits its weapon programs and offices would also be one that systemically uncovers incompetent and crooked managers, false promises and those who made them. It would also necessarily reveal reasons to dramatically alter, if not cease, funding for some programs, which of course would make lots of people in industry, Congress, and the executive branch unhappy.

The current system and its out of control finances mortally harm our defenses, defraud taxpayers, and bloat the Pentagon and federal budgets. Any reform that fails to address this most fundamental problem is merely another doomed attempt that will only serve to perpetuate a system that thrives on falsehoods and deception.

Essay 9

“Evaluating Weapons: Sorting the Good from the Bad”

by Pierre M. Sprey

The world is awash in mediocre or even useless weapons. The good ones are few and far between. Telling the difference is of utmost consequence to the people who have to use the weapons—and to the nation that has to pay for them.

If you are seriously trying to understand whether a given fighter, destroyer, tank, rifle or truck is worth acquiring, you will soon find yourself buried under a mountain of misinformation—the more expensive the weapon, the deeper you’ll be buried. Here are a few guideposts for digging your way out:

RULE 1: *Weapons are* not *the most important ingredient in winning wars. People come first; ideas are second and hardware is only third.*

After 1973’s crushing 80-to-1 victory by Israelis flying F-4s and Mirages against Arab pilots flying MiGs, the commander of the Israeli Air Force (IAF), Gen. Mordecai Hod, famously remarked that the outcome would have been the same if both sides had swapped planes. He was exactly correct, simply because the IAF had the most rigorous system in the world for filtering out all but the most gifted pilots. In every war, it’s the few superb pilots that win the air battle. A tiny handful of such pilots have dominated every air-to-air battleground since World War I: roughly 10 percent of all pilots (the “hawks”) score 60 percent to 80 percent of the dogfight kills; the other 90 percent of pilots (“doves”) are the fodder for the hawks of the opposing side.[1] Technical performance differences between opposing fighter planes pale in comparison.

---

[1] See Herbert K. Weiss, “Systems Analysis Problems of Limited War,” *Annals of Reliability and Maintainability*, AIAA, New York, July 18, 1966. Weiss’ extensive probing of air combat, submarine and land battle data are among the most original and useful quantitative analyses of combat data ever done. Available at http://pogoarchives.org/labyrinth/09/01.pdf.

Submarine warfare is strikingly similar: the best 10 percent of the skippers account for the majority of the tonnage sunk. And, when the ace skippers switch boats, the high scores go with the skipper, not with the crew left behind.

Ground combat is much subtler and more complex than air or naval warfare—thus, relative to hardware, people and ideas are even more dominant. In 1940, the Germans, outnumbered 1.5 to 1 in armor by French and British tanks[2]—most of them technically superior—crushed France in just three weeks. The smaller German tank forces hardly mattered; they won because they had far better combat leaders, tactics and morale, and because their troops were far better trained. Fifty years later, commenting on a similar disparity in people, General Schwarzkopf said the outcome of Gulf War I would have been the same if the U.S. and Iraqi armies had exchanged weapons—thereby echoing General Hod.

People are so overwhelmingly important in war that, as we shall see in Rule 5, the single most important characteristic of a weapon is its effect on the user, that is, whether it helps or hurts the user's combat skills, adaptability and fearlessness.

RULE 2: *Not all weapons are equally important in war. Their importance is unrelated to their cost.*

Rifles and machineguns, cheap as they are, are far more important than fighters or bombers in winning wars. That's as true today as it was in World War II. As thoughtful observers have noted, the ubiquitous availability since the 1950s of automatic (burst fire) rifles like the AK-47—as opposed to previous semi-automatic (single shot per trigger squeeze) rifles—is a dominant leveling factor in the astonishing success rate of guerillas against much better equipped regular armies over the last half century. As just one example, in small unit firefights early in the Vietnam War, the AK-47-equipped Viet Cong irregulars had a significant exchange rate advantage over U.S. infantry, despite huge U.S. advantages in artillery, helicopters, radios and vehicles. Sadly, the U.S. infantryman was much hampered by his M-14, a heavy and cumbersome rifle, entirely unusable when in burst fire mode.

That is exactly why in 1963 the theater commander, General Westmoreland, reviewing the remarkable firefight successes of units combat testing a

---

[2] Karl-Heinz Frieser, *The Blitzkrieg Legend* (Naval Institute Press, 2005). Frieser reports 3,554 British-French tanks (including 300 British) and 2,429 German ones; in total vehicles, the Allies had 300,000 versus only 120,000 for the Germans, still heavily reliant on horse-drawn transport. The crucial German tank advantage was in the "people" domain: *each tank had a radio.* Allied tanks had essentially no radios.

remarkably light and reliable new automatic rifle, the commercially-produced AR-15, immediately demanded that the AR-15 replace the M-14 throughout Vietnam—over the violent objections of the entire U.S. Army ordnance bureaucracy, all die-hard defenders of the M-14 they had spawned. Fearing Army-wide replacement of their pet, the small arms bureaucrats delivered to Westmoreland in late 1964 a "militarized," heavier, less effective version of the AR-15, the infamous early M-16A1, which they deliberately furnished with a powder that would make it jam in combat.[3] As a result, young GIs died with jammed M-16s in their hands. It took three years and a brutally incisive congressional investigation[4] to force the Army bureaucracy to fix the M-16 they had sabotaged.

Other examples of crucially important, cheap—and therefore neglected—systems spring quickly to mind. Acquiring a better five ton truck has far more impact than C-5 or C-17 airlifters on the mobility and sustenance of our troops in battle—but doesn't receive one-hundredth as much congressional or public attention. Similarly, our troops have no squad radio that is effective in jungles, woods and cities. Such a $250 walkie-talkie would do more for winning firefights and saving GI lives than the elaborate, $15 billion JTRS digital do-everything command and control radio network that is the Defense Department's current infatuation.

Weighing the results of the last 70 years of air warfare, cheap $15 million close air support planes will clearly contribute far more to saving American troops in trouble and to winning wars than $2.2 billion B-2s or $160-plus million "multipurpose" fighters like the F-35[5]—no matter whether we're facing Taliban fighters or massed tanks.

---

[3] Col. Richard R. Hallock, (U.S. Army, ret.), "M-16 Rifle Case Study," March 16, 1970. (Prepared for the chairman of the President's Blue Ribbon Defense Panel.) This is a document of historic significance, not previously available: a uniquely accurate, insightful, objective and carefully documented account of the M-16's development and the malign battle of the Army bureaucracy—up through the chief of staff—to prevent its adoption. Written by an insider who was an eyewitness to the entire tragedy, from the rifle's brilliant genesis through a searing congressional investigation of Army culpability. Find a copy of this document at http://pogoarchives.org/labyrinth/09/02.pdf.

[4] "Report of the Special Committee on the M16 Rifle Program of the Armed Services Committee of the House of Representatives," October 19, 1967. The Ichord Report stands as one of the all-too-few landmarks of incisive congressional oversight, a must-read for anyone who wants to understand how and why entrenched and incompetent weapons acquisition bureaucrats supported by sleazy contractors lead directly to deaths in combat. Find a copy of this document at http://www.vietnam.ttu.edu/star/images/256/2560131001a.pdf for the first 50 pages and at http://www.vietnam.ttu.edu/star/images/256/2560131001b.pdf for the last six.

[5] See Pierre M. Sprey, "Notes on Close Air Support," Intrec Inc. Internal Study, Potomac, MD, May 1974. This is an extended introduction to the nature of the close air support

Victory at sea is equally unrelated to weapons cost. By the end of 1914, 28 diminutive German submarines, each one-fortieth the cost of a battleship, had wrested control of the seas from the 47 mighty battleships, 195 cruisers and 200 destroyers of the Royal Navy. The battleship had become irrelevant forever—though the obstinacy of hidebound admirals and the corrupting power of lucrative procurement budgets kept the battleship in full tilt production for 30 more years.

And in its carrier reincarnation, the battleship is still soaking up the lion's share of the U.S. Navy budget to this day. The preoccupation with $14 billion carriers escorted by $1 to $3 billion destroyers has led to virtually complete Navy neglect of strategically essential coast control capabilities like $175 million minesweepers, $60 million coastal patrol ships, $35 million fast missile-torpedo boats and $4 million riverine-estuarine warfare boats. In the 1991 Gulf War, the Navy's perennially inadequate minesweeping forces made it too dangerous to launch a 17,000 Marine amphibious assault that General Schwarzkopf had planned.[6] Recently, in the Indian Ocean, the U.S. Navy's utter lack of coastal patrol and fast attack boats left our merchant ships mostly unprotected against pirates in rubber skiffs. As a result, we witnessed the ludicrous scene of using a $1 billion destroyer to subdue four rifle-armed pirates in a 25-foot inflatable.

RULE 3: *You can't tell effective weapons from useless ones without a clear definition of each combat-essential effectiveness characteristic—and that definition must be derived directly from combat evidence.*

Consider the marksman's definition of rifle effectiveness: the ability to kill a standing soldier at 500 yards with one shot. That's plausible to the layman but laughably irrelevant to anyone who's ever been in an infantry firefight. Pursue the marksman's definition and you'll pick a rifle that's got so much recoil, is so heavy and puts out so few rounds that it's nearly useless to the average 19-year-

---

(CAS) mission, the effectiveness characteristics required, and a comparison of aircraft available for the mission in 1974 (which remains essentially unchanged today, since no new CAS-specific aircraft or weapons have been developed in the intervening 35 years). Find a copy of this document at http://pogoarchives.org/labyrinth/09/03.pdf. See also Pierre M. Sprey, "Combat Effectiveness Considerations in Designing Close Support Fighters," Briefing for the Office of the Secretary of Defense and for the Industrial College of the Armed Forces, 1983. This includes an effectiveness analysis, design characteristics and cost for a feasible close air support aircraft significantly more lethal and survivable than the A-10 at one-fourth the size and half the cost. See this document at http://pogoarchives.org/labyrinth/09/04.pdf.

[6] Marvin Pokrant, *Desert Storm at Sea: What the Navy Really Did* (Westport: Greenwood, 1999), 98.

old GI ambushed by insurgents spraying lethal bursts from ancient but fully automatic AK-47s.

In stark contrast to the marksman's dream, real infantry rifle combat occurs far more often at 15 to 50 yards than at 500—and never involves single shots or single shooters. Targets are rarely more visible than a momentary muzzle flash or puff of smoke. Getting lots of rounds off nearly instantly is of overwhelming importance. Near misses (suppressive fire) are almost as useful tactically as hits. For a brief exposition of how this distillation of actual rifle combat translates into quantitative effectiveness measures, see below.[7]

Similarly, real air-to-air combat is separated by a chasm from the technologist's dangerously beguiling dream of beyond-visual-range (BVR) combat: push a button, launch a missile at a blip on the scope at 25 miles, then watch the blip disappear without ever having laid eyes on the target. That concept of combat, oblivious to the inconvenient details of real air-to-air fights[8], leads to huge, cumbersome fighters loaded down with tons and tons of heavy stealth skins, massive radars and missiles, and failure-ridden electronics of unmanageable complexity. The most recent fighter built in pursuit of the BVR combat delusion, the F-22, has a $355 million sticker price and costs $47,000 per hour to fly, making it impossible to fly the hours necessary to train pilots adequately (people first!)—and impossible to buy enough fighters to influence any seriously contested air war.

As opposed to the BVR dream, actual air combat almost invariably starts with two or more attackers "bouncing" and surprising an unaware flight of fighters at

---

[7] Pierre M. Sprey, "Coming to Grips with Effectiveness in Rifles," Informal Briefing for the Office of the Secretary of Defense, 1981 and for the Congressional Military Reform Caucus. Presents a very brief synopsis behind the brilliant measures of rifle effectiveness developed and defined by Col. Richard R. Hallock as a basis for his 1965-1966 CDEC Small Arms Weapon System (SAWS) Field Experiment. Find this at http://pogoarchives.org/labyrinth/09/05.pdf. For a more detailed, formal definition of these measures and the associated test conditions, see pp. III-3 to III-8 in "The Evaluation of Small Arms Effectiveness Criteria, Volume 1," Intrec Inc. for Defense Advanced Research Projects Agency, May 1975. This is the clearest available description of the pioneering SAWS Field Experiment, including the meticulously realistic details of the computerized target ranges, the training of test subjects, the squad firing scenarios and the extraordinary measures for preserving the test's all-important target range unfamiliarity. Find a copy of this document at http://pogoarchives.org/labyrinth/09/13.pdf.

[8] See Lt Col Patrick Higby, U.S. Air Force, "Promise and Reality: Beyond Visual Range (BVR) Air-to-Air Combat," Research Paper prepared for Air War College Electives Program, Maxwell Air Force Base, March 30, 2005. This paper is available at http://www.vmi.edu/uploadedfiles/archives/adams_center/essaycontest/20042005/higbyp_0405.pdf. It is also available at http://pogoarchives.org/labyrinth/09/06.pdf.

their normal cruise speed (no more than mach .7 to .9 for all existing fighters). The surprise factor looms large: in every war of the past century, 75 percent to 90 percent of all pilots shot down in air-to-air combat were unaware. Attackers must close to within roughly a quarter mile or less to get positive eyeball identification of friend or foe (no current electronic identification is secure enough to prevent shooting friends) before maneuvering into missile or cannon firing position, then getting a shot off as quickly as possible. If the defenders wake up (an infrequent occurrence among "doves"), or if the attackers' first firing pass misses (a frequent occurrence), a dogfight ensues with both sides maneuvering to gain firing position and to defeat enemy firing passes.

To win this kind of fight places a premium on gifted pilots, above all else. In distant second place are the airplane characteristics that will help those pilots to win, as follows:

- achieving surprise by visual and electronic undetectability, e.g. tiny size, no radar emissions and higher *cruise* speed than the enemy's (which ensures that he can't sneak up from behind);

- ability to launch lots of friendly fighters into enemy skies every day (achieved through low sticker price, low maintenance leading to many sorties per day and long cruise endurance) and ability to generate lots of air combat training hours (ditto) to produce plenty of gifted pilots;

- superior agility—i.e., better turn, better acceleration and quicker control response—to gain firing position and defeat enemy firing passes (less weight, more thrust and more wing area each increase agility);

- carrying weapons that deliver reliable kills *quickly* (cannons first, simple infrared missiles second, radar missiles are off the table since they are neither quick nor reliable).

For a more thorough treatment of real fighter combat, and how it shapes effectiveness characteristics, see below.[9]

---

[9] Pierre M. Sprey, "Comparing a Quarter Century of Fighters," Straus Military Reform Project, Center for Defense Information. April 2006, http://www.cdi.org/pdfs/Sprey%20Quarter%20Century.pdf. The briefing introduces combat-derived measures of effectiveness for air-to-air fighters, measures that are then used to compare existing fighters. See also Pierre M. Sprey, "Comparing the Effectiveness of Air-to-Air Fighters: F-86 to F-18." The study, available at http://pogoarchives.org/labyrinth/09/08.pdf was released by the Office of the Secretary of Defense (PA&E) in April 1982. It defines measures of effectiveness in detail for air-to-air fighters based on combat data, evaluates the effectiveness of past and contemporary

In a similar vein, studying the great successes achieved by tank forces in combat quickly dispels the two pillars of orthodox armor wisdom: first, that combat judges tanks by how well they fight other tanks and, secondly, that the cannon is the tank's most important weapon. Neither dogma has anything to do with the way George Patton or Heinz Guderian employed armor in achieving their astonishing victories. For a more realistic view of tank combat and a definition of tank effectiveness that is more useful in weeding out bad tanks and designing better ones, see a briefing prepared by this author in 1979.[10]

RULE 4: *To understand the characteristics that separate weapons effective in combat from mediocre or useless ones, read ten times more combat histories than research and development (R&D) sagas or weapons technology eulogies. Most useful are combat histories from the foxhole, cockpit or periscope point of view.*

One read through pioneering combat historian S.L.A. Marshall's "Men Against Fire"[11] will teach you more about how rifles are used in combat—and the huge edge enjoyed by burst fire over single shots—than two trailer truckloads of U.S. Army Materiel Command rifle analyses. His 1958 "Sinai Victory"[12] chronicles how raggedy-looking but superbly-trained Israeli platoon leaders and troops, using ancient World War II .50 caliber-equipped jeeps and hand-me-down Spitfire aircraft, achieved blitzkrieg results that none of their contemporary tank-and-jet equipped armies would have been able to match.

---

fighters from around the world, and then, using the same measures, synthesizes the design characteristics of a new ultra-agile, ultra-small supercruising fighter (of demonstrably higher effectiveness than today's F-22). For a discussion of effectiveness across several types of weapons using these combat derived criteria, see http://pogoarchives.org/labyrinth/09/07.pdf.

[10] Pierre M. Sprey, "Comparing the Effectiveness of Current Tanks," Briefing for Office of the Secretary of Defense, 1979. Derives combat-history-based measures of effectiveness for tanks and compares the M-1, the M-60 and the T-62. Find a copy of this document at http://pogoarchives.org/labyrinth/09/10.pdf.

[11] S.L.A. Marshall, *Men Against Fire: The Problem of Battle Command* (New York: Morrow, 1947). This is a path-breaking analysis of when and why soldiers do or don't fight. Also read the essential follow-on, Marshall's *The Soldier's Load and Mobility of a Nation* (published by the Marine Corps Association and others) on the rapid destruction of fighting spirit when the infantryman's load exceeds 40 pounds—a central though widely ignored constraint when designing small arms, anti-tank weapons or any other infantry equipment.

[12] S.L.A. Marshall, *Sinai Victory* (New York: Morrow, 1958). Uses the 100 hour Israeli campaign of 1956 to paint an unparalleled picture, rich in combat detail, of why people are vastly more important than hardware. It contains a must-read appendix on the eminently sensible Israeli methods of training for lightning tactical decisions under combat stress.

Read Japanese World War II ace Saburo Sakai's "Samurai!"[13] and Wing Commander H.R. Allen's "Who Won The Battle of Britain"[14] and you'll know far more about the realities of air combat than if you had absorbed every official U.S. Air Force history from World War II to Desert Storm.

To come to grips with the essence of submarine warfare, start with "Silent Victory" by Clay Blair Jr.[15] If you want to understand fast attack boat combat and how much relevance the Navy has lost by neglecting it, read "PT-105" by Dick Keresey[16] and "The Battle of the Torpedo Boats" by Bryan Cooper.[17]

RULE 5: *For any weapon, the list of essential effectiveness characteristics must include the weapon's direct effect on the user's skill, combat adaptability and training (people first!)—and, equally important, the effect on the number of weapons (i.e. the force level) actually delivered on the battlefield. Any definition of effectiveness lacking these two elements is useless.*

In rifles, the effect of the weapon on the user's skill is all too obvious: the four-fold reduction in "kick" (i.e., recoil energy) of the 5.56 mm bullet of the M-16 versus the 7.62 mm of the M-14 allows the average infantryman to put more

---

[13] Saburo Sakai, Martin Caiden and Fred Saito, *Samurai!* (Bantam, 1985). The great Japanese ace's superb insights into the dominance of pilot ability, the gulf between the gifted and ungifted pilot, and how the United States achieved air superiority, not by bombing fighter factories but by decimating Japan's gifted pilots in the air.
[14] Wing Commander H.R. Allen, *Who Won the Battle of Britain?* (London: Barker, 1974). This is a common sense, eyewitness account of how inept tactics and appalling Royal Air Force command incompetence caused needless slaughter of young British fighter pilots while allowing the Luftwaffe to gain air superiority over England for two weeks.
[15] Clay Blair Jr., *Silent Victory: The U.S. Submarine War Against Japan* (Naval Institute Press, 1975). This meticulously researched history drives home the dominance of the submarine in the strangling of the Japanese economy, as well as the huge gap in combat results between good and bad skippers. It is commendably frank on the many inexcusable U.S. Navy command blunders: the admirals' short shrift for submarines before and during the war, their incompetent torpedo procurement, their combat-irrelevant tactical doctrine, and their grossly inadequate training and selection of skippers.
[16] Dick Keresey, *PT-105* (U.S. Naval Institute Press, 1996). Chronicles the disproportionate contributions of the lowly, cheap fast boat in interdicting Japanese army transports as well as Imperial Navy fighting ships. It drives home the overwhelming importance of controlling coastal waters and the futility of trying to do so with a deepwater Navy.
[17] Bryan Cooper, *The Battle of the Torpedo Boats* (London: MacDonald, 1970). Covers the strategic importance of fast boat coastal operations and their interdiction successes in the D-Day, Dunkirk, North African and Italian campaigns as well as in the Aegean—and, of course, the fast boat's major role in the Southwest Pacific island-hopping strategy.

bullets on or near the target at any combat-relevant range (and with less training), as is convincingly demonstrated by several critically important analyses of rifle field tests.[18]

In fighters, the effect of high cost and the associated burden of high maintenance downtime are equally obvious. The F-22 costs 10 times as much as an early model F-16 fighter and, due to its huge maintenance load, can fly only half as many sorties per day. Thus, for equal investment, the F-22 delivers only one-twentieth as many airplanes over enemy territory as the F-16—a crippling disadvantage, no matter whether the F-22's stealth and weapons work or don't work.

RULE 6: *In sorting good weapons from bad, relying on R&D test results for assessing combat accuracy, probability of kill, reliability, effective range, etc. is disastrous. Sadly, operational or field test results have become almost equally useless, except for occasionally uncovering unanticipated problems. Unfiltered, non-anecdotal samples of combat results trump everything else.*

Though vastly harder to implement than any outsider can conceive, honest and realistic effectiveness testing of weapons is feasible. But the inherent military bureaucratic obstacles have grown so insurmountable that I know only two examples of truly combat-representative testing, uninfluenced by the procurement bureaucracy: the uniquely brilliant and realistic 1965-1966 SAWS M-14 vs. M-16 vs. AK-47 field test[19] and the A-10 Armament Directorate's Lot Acceptance Verification Program (LAVP) for 30 mm rounds,[20] a superb 1978 airborne firing lethality test against 300 fully functional Soviet and U.S. tank targets that inspired the Live Fire Testing Program mandated by the Congress. Since 1978 there have been essentially no similarly realistic effectiveness tests.

---

[18] See pp. 43-59 of "Small Arms Weapon System Analysis: A Review and Evaluation," Department of the Army, Office of the Chief of Staff, 1966. Not previously available, this insightful appendix of a larger study was commissioned, read and then ignored at the highest levels in the Army. Find a copy of this document at http://pogoarchives.org/labyrinth/09/09.pdf. See also pp. 88-90 of the "M-16 Rifle Case Study " (footnote 4) and "Coming to Grips with Effectiveness in Rifles" (footnote 7).

[19] See pp. IV-1 to IV-46 of "The Evaluation of Small Arms Effectiveness Criteria, Volume 1," Intrec, Inc. for Defense Advanced Research Projects Agency, May 1975, available at http://pogoarchives.org/labyrinth/09/13.pdf.

[20] Pierre M. Sprey, "The Terrible Cost of Not Testing with Real Weapons Shooting at Real Targets," Briefing presented to the U.S. Air Force Armament Development and Test Center (Eglin Air Force Base) and to the Congressional Military Reform Caucus, 1979. Contains useful insights into the early roots of live fire testing in the DOD and examples of the tragic combat consequences of flawed testing. Find a copy of this document at http://pogoarchives.org/labyrinth/09/11.pdf.

R&D tests, though perhaps useful to designers and engineers, are inherently useless for judging a weapon's effectiveness because they suffer from an insurmountable conflict of interest: they are controlled by the weapon's development agency. Developer agencies always have a powerful vested interest in proving that their creation is a success and more effective than any alternatives. In theory, operational or field tests, that is, those run not by developers but by military end-users, are free of this conflict. In reality, the "keep the money flowing" pressures of contemporary military senior leadership make rigorous, honest and useful user tests impossible. A 1981 briefing catalogs the most common—and still ongoing—abuses in operational testing.[21] In the nearly three decades since, the list of OT&E abuses has hardly changed, though the bias in test outcomes has become far more egregious. The single most crippling new abuse is the now-common practice of having contractors (or their subsidiaries) "participate" in the writing of operational test reports evaluating their own product.

A dramatic example of the gulf between the rosy optimism of R&D testing and the brutal reality of combat is the AIM-7 Sparrow air-to-air radar missile, the mainstay of the technologists' hopes of beyond-visual-range combat for at least 40 years. The Sparrow's initial R&D tests reported 80 percent to 90 percent kill rates. Of course, nearly 100 percent of these tests were against non-maneuvering drone targets, many of them with artificially strengthened radar returns. Operational tests claimed 50 percent to 60 percent kill rates, shooting at mostly non-maneuvering targets with a token light maneuver thrown in now and then.

Combat reality raised its ugly head in the skies over North Vietnam. Successive "improved" Sparrow models from the AIM-7B to the AIM-7F never got above the 8 percent to 10 percent hit rate. Lots of angry F-4 fighter pilots came home cursing about getting a perfect tail position on a MiG, firing all four Sparrows on board, and watching all four miss. And, bitterest pill of all, they had no cannon onboard the F-4B/C/Ds to use after the missiles missed. Ironically, the Sparrow's highly touted 90 percent R&D kill rate was the aircraft bureaucracy's prime excuse for omitting the gun.

Combat proved the AIM-7 to be worse than useless: the drag and weight penalties of carrying four large missiles and of the expanded fuselage needed to hold the large, heavy radar and its bulbous radome sorely degraded the dogfighting performance of the F-4—as well as that of the later F-14, F-15 and F-18.

---

[21] Pierre M. Sprey, "Today's OT&E: Abuses and Remedies," Informal Briefing by Pierre M. Sprey for Congressional Military Reform Caucus, 1981. Find a copy of this document at http://pogoarchives.org/labyrinth/09/12.pdf.

Similar glowing peacetime test reports followed by fatal combat failures can be reported for a multitude of other systems. The $1 million per shot Tomahawk cruise missile passed its Navy Operational Evaluation tests with flying colors. In Gulf War I, DOD admitted the Tomahawk failed to fly and find the target nearly half the time; the true effectiveness rate in attacking actual targets was classified, but lower.[22] Five different expensive radar jamming pods—the ALQ-75, 76, 77, 81 and 87—all passed their operational tests and were sent into Vietnam combat to protect fighters against radar surface-to-air missiles. All five failed. To the end of the war, pilots had to defeat missiles by outmaneuvering them, often while burdened with the heavy pods.

As final food for thought, the testing morass has serious implications for the nation's imagined strategic nuclear capabilities. The accuracy and reliability of our ICBMs are tested under the same appallingly unrealistic conditions and the same "keep the money flowing" pressures as our air-to-air missiles. As a result, it is entirely conceivable that the wartime launch reliability of ballistic missiles and their target miss distances could be *an order of magnitude* worse than reported to the President and to our highest military commanders.

RULE 7: *When judging weapons effectiveness, seek out informed skeptics, both in and out of uniform. Weigh carefully their insights on weapons shortcomings. Ignore the corporate flacks, military procurement program managers, acquisition command flag officers, civilian high tech advocates and, above all, the "experts" and "experienced users" trotted out by the military services whenever their favorite programs are under attack.*

No example demonstrates better the enormous value of an informed skeptic than the Patriot tactical ballistic missile defense system. During Gulf War I, 158 Patriots were fired at incoming Iraqi Scud ballistic missiles, an ancient and ineffective derivative of the World War II German V-2 rocket. Army press releases during the war claimed 100 percent of Scuds were shot down, reducing this to 96 percent in the first testimony to Congress, then 80 percent, 70 percent and a final figure of 52 percent, though with a caveat that only 25 percent could be supported with "high confidence." The Army's slow backpedaling from their initial outrageous claims was entirely due to the meticulous analyses of combat videotapes by a single courageous, highly qualified skeptic, M.I.T. professor Theodore Postol. His final work demonstrated that, *at best*, only 2 to 4 of the 158 incoming Scuds had been destroyed by Patriots, even though more than 3 Patriots were fired at each Scud, on average. In truth, Postol showed there was no conclusive evidence that *any* Scuds had been destroyed by Patriots.

---

[22] See p. 141 of "Operation Desert Storm: Evaluation of the Air Campaign," U.S. General Accounting Office, GAO/NSIAD-97-134, June 1997, http://www.gao.gov/archive/1997/ns97134.pdf.

Even worse, when the Patriots were deployed to defend Tel Aviv halfway through the Iraqi Scud campaign, Postol's evidence showed they *increased* Israeli casualties per Scud by 74 percent and apartments damaged per Scud by 340 percent—apparently mostly due to explosion debris from the large numbers of Patriots that missed.[23]

Needless to say, the 0 percent to 5 percent combat success rate of Patriot batteries against the primitive Scuds is a poster child for the false claims and likely failures in combat of our $90 billion Ballistic Missile Defense System.

Wrap-Up

There can be no question that independent, reasoned, combat-based effectiveness assessments of our major weapons programs by people both inside and outside DOD are needed more than ever. Be under no illusions about the huge obstacles facing any such attempts—obstacles imposed by corporate hunger for profits, by encrusted military procurement bureaucracies pursuing their self-interest and by military users slavishly defending traditional doctrine. Tackling these powerful interests takes guts and tenacity. But if we don't take them on, the country will continue to pay more and more for shrinking forces that contribute less and less to our nation's security.

[23] Other accounts of the non-success of Patriots in the First Gulf War may vary regarding the details, but they all agree on the fundamental message. Hearings in Congress in April 1992 left serious doubt whether any Scuds had been effectively hit by Patriots. See the testimony for these hearings at http://www.fas.org/spp/starwars/congress/1992_h/. For further analysis, also see George N. Lewis, "How the US Army Assessed as Successful a Missile Defense that Failed Completely," *Breakthroughs of the Security Studies Program of MIT* 12, no. 1 (Spring 2003). http://web.mit.edu/ssp/publications/breakthroughs/Breakthroughs03.pdf. .

Essay 10

"Developing, Buying and Fielding Superior Weapon Systems"

by Thomas Christie

The current Defense Department acquisition process that develops, tests and procures new weapons for U.S. combat forces has evolved over the past five decades in response to multiple defense management strategy initiatives, external reform proposals and lessons-learned from the field. The conventional wisdom notwithstanding, the process as spelled out in DOD's directives and instructions is fundamentally sound and could avoid its unending cost overruns, delays and performance failures, if it were implemented in a better informed and rigorously disciplined manner. The problem is not nearly as much in the laws and regulations as it is in the execution by the people who have been operating the system.

We should not waste time in this short essay reinventing bromides for the bureaucracy to cogitate and self-appointed reformers to contrive. Essential ingredients to a viable weapons acquisition system include –

- budgeting with truly independent estimates of program development, procurement and support costs;
- an evaluation process, again independent, to find and correct reliability problems early and throughout the entirety of a program's life cycle, and
- conducting combat realistic operational tests of weapons and honest and complete reports to permit decision-makers inside and outside the Pentagon to make properly informed judgments.

Anyone paying attention to the way the system has broken down up to now knows these are needed, but there is also more. There are other features of the process that need attention and must be executed, not circumvented, to achieve successful weapons at affordable cost in a reasonable time. These other essential aspects include –

- insisting on discipline throughout the decision-making process;
- cleaning up the front end of the process where dubious requirements and buy-in cost and schedule promises are greeted without criticism and committed to;
- demonstrating—through empirical field testing, not success-oriented modeling and simulation—new technologies before each major decision-maker approval point;
- establishing and carrying out event-based strategies accompanied by realistic pass/fail criteria for each phase of a program;
- conducting continuous independent evaluations of programs all the way through development, testing, production, and even after introduction in the field—to include training exercises and combat results, and
- feeding back all such results completely and accurately to the entire acquisition community.

Nothing in today's laws and regulations prevent any of the above; most are actually called for, and yet they almost never happen.

The Need for Reform Is Not New

Proceeding with any new weapon system development, production and fielding with the Pentagon acquisition process as currently implemented (or, perhaps more appropriately, not implemented) will only continue the debacles of the past. Both past and present Pentagon leadership has been painfully aware that "Something's wrong with the system," as Secretary of Defense Donald Rumsfeld told Congress in 2005.[1]

---

[1] See Michael Bruno, "'Bow Wave' Of Acquisition Costs Coming," Rumsfeld Tells Senators," *Aviation Week*, April 28, 2005. Further evidence of Rumsfeld's concern came in a June 7, 2005 memorandum from his acting deputy secretary of defense, Gordon England. Addressed to senior Pentagon leadership, it directed a thorough assessment of the acquisition process "to consider every aspect of acquisition, including requirements, organization, legal foundations, decision methodology, oversight, checks and balances – every aspect." In kicking off yet another study at the time, England stated: "Many programs continue to increase in cost and schedule even after multiple studies and recommendations that span the past 15 years." (See England's "Acquisition Action Plan," June 7, 2005, described at http://www.heritage.org/Research/Reports/2005/10/Congressional-Restraint-Is-Key-to-

More recently, Secretary of Defense Robert Gates was perceptive in stating –

> "First, this department must consistently demonstrate the commitment and leadership to stop programs that significantly exceed their budget or which spend limited tax dollars to buy more capability than the nation needs…
>
> Second, we must ensure that requirements are reasonable and technology is adequately mature to allow the department to successfully execute the programs…
>
> Third, realistically estimate program costs, provide budget stability for the programs we initiate, adequately staff the government acquisition team, and provide disciplined and constant oversight.
>
> We must constantly guard against so-called "requirements creep," validate the maturity of technology at milestones, fund programs to independent cost estimates, and demand stricter contract terms and conditions."[2]

There is nothing wrong with the assertions, but even with Secretary Gates' many subsequent program alterations, a few actual cancellations, and some modest overhead savings, can anyone say that the Pentagon has transformed into what Gates said he wanted? More, much more, actual implementation is required.

Congress has behaved similarly—with words more grandiose than actions. In 2009, it weighed in with its latest attempt to rescue the Pentagon's acquisition processes: the Weapon Systems Acquisition Reform Act of 2009 (WSARA 2009). In addition to re-establishing test and evaluation and system engineering capabilities eliminated by the Clinton administration with Congress' consent, WSARA 2009 directed the application of several ideas that had been advocated for decades; these included independent cost assessments; evaluating trade-offs of cost, schedule and performance; and competitive prototype development and testing.

But will the Pentagon actually follow what Congress says it intends with this legislation, or will it exercise the many loopholes that Congress consciously inserted into virtually every requirement—at the explicit request of top DOD

Successful-Defense-Acquisition-Reform and available at http://www.thefreelibrary.com/Subject%3a+Acquisition+Action+Plan.-a0140554367.

[2] Gates made these comments in his Defense Budget Recommendation Statement on April 6, 2009, available at http://www.defense.gov/speeches/speech.aspx?speechid=1341.

management—to permit circumvention of most, or all, of these reforms? History suggests the latter.

The Problem is Not Lack of Study

It is difficult to find another process that has been studied more than DOD acquisition. Every three to four years, yet another high-level study is commissioned to review DOD management in general and the acquisition process in particular, or Congress steps in and legislates, in great detail, how the Pentagon should organize and carry out its mission. The commissions, studies and statutes are many.[3]

The common goal for many of these efforts has been "streamlining" the acquisition process. Typical techniques recommended were efforts, not always successful, to reduce management layers, eliminating reporting requirements, replacing regulated procurement with commercial off the shelf (COTS) purchasing, reducing oversight from within as well as from outside DOD, and eliminating perceived duplication of testing.

Advertised as reform, most of these efforts had the real effect of reducing objective supervision and independent management. While oversight by government agencies and the associated reporting requirements can indeed be burdensome, they are not the causes of the continuing miserable record of program stretch-outs and cost growth. This is true independent of whether those agencies and their reporting requirements are internal to DOD, such as the Defense Contract Management Agency (DCMA), independent cost analysis groups, and operational test and evaluation organizations; or external entities, such as the congressional committees and the Government Accountability Office (GAO). This finding is borne out by my decades of participation in the acquisition process and some of the more competent official reviews, such as that done in 1990 by the Defense Science Board (DSB). [4]

---

[3] The more recent ones include but are not limited to the following: The 1970 Fitzhugh Blue Ribbon Commission, the 1977 Steadman Review, the 1981 Carlucci Acquisition Initiatives, the 1986 Packard Commission and Goldwater-Nichols Act, the 1989 Defense Management Review, the 1990 Defense Science Board (DSB) Streamlining Study and another DSB Acquisition Streamlining Task Force in 1993-1994, the Total System Performance Responsibility (TSPR) initiative of the late 1990s, the early 2000s focus on Spiral Development and Capabilities-Based Acquisition, the Defense Acquisition Performance Assessment (DAPA) of 2006, the DSB Task Force of 2008 on Development Test and Evaluation, the 2009 Weapon System Acquisition Reform Act, and the new "IMPROVE" Acquisition Act passed by the House of Representatives in 2010.

[4] This DSB Task Force on Acquisition Streamlining was commissioned by the Under Secretary of Defense for Acquisition, John Betti, in late 1989 and was chaired by John Rittenhouse, a General Electric corporative executive. A sub-group of that task force

That 1990 DSB review concluded that failure to identify and admit to technical problems, as well as real costs, before entry into what was known as Full-Scale Engineering Development (FSED)—now referred to as Engineering and Manufacturing Development (EMD)—was the overwhelming cause for subsequent schedule delays, often years, and the resulting cost growth. Oversight enabled the discovery and reporting of test failures during FSED/EMD that often necessitated additional time and dollars for system redesign, testing and retesting of fixes, and costly retrofits of those fixes. It is a viable question, however, whether these delays discovered early caused more, or less, schedule alteration to utility in the field than discovering the problems late, after deployment. Without question, testing and finding problems early, before serial production, is cheaper – by a very large margin.

After all the reforms of previous decades, here we are in 2010 and what's demonstrably different from the past? Major defense programs are taking 20 to 30 years to deliver less capability than planned, very often at two to three times the cost. It all may be worse now than ever before.

The basic problem is not the official directives.[5] Instead, Pentagon acquisition officials too often have violated the regulation's intent by approving "low-balled" estimates of the costs and time required to deliver new capabilities, and ignoring independent assessments that were often available and more realistic. Time and again, early-on funding for building and testing prototypes to better understand technical and operational issues has gone by the wayside. A powerful – overwhelming – factor in the making of these slipshod decisions is the competition for dollars inside the bureaucracy: approve the money now, lest it be grabbed by another program.

A typical hardware program will involve three to five administrations and ten, or more, congresses. By the time the technical and cost issues finally become known, few, if any, of those involved initially are still around, and those who are refuse to admit they had been wrong, to cut their losses before the problems

---

examined some 100 major programs under OSD oversight during the 1986-1990 timeframe. Most of the programs were plagued by cost increases and schedule stretch-outs; the study group used available program documentation and extensive interviews with DOD officials to determine root causes for these problems. A final DSB report was never published, but the Institute for Defense Analyses produced IDA Paper P-2551, entitled "The Role of the Office of the Secretary of Defense in the Defense Acquisition Process," documenting the sub-group's analyses and findings. It is available at http://pogoarchives.org/labyrinth/10/01.pdf.

[5] Find these materials, DOD Directive 5000.1 and DOD Instruction 5000.2 governing the Pentagon's acquisition process, at http://www.dtic.mil/whs/directives/corres/pdf/500001p.pdf, and http://www.dtic.mil/whs/directives/corres/pdf/500002p.pdf.

worsen, or to discipline the system by making an example of program officials and their contractors who have sold the department and the taxpayers a bill of goods.

To be fair, there are indications more recently a Pentagon leader has begun to take these considerations to heart in his decision-making. Secretary Gates has stopped further production on one major program (the F-22); he has reduced the future buy for others (such as the DDG-1000), and he has reconstituted several under new nomenclature, requiring a redo (such as the Future Combat Systems and the VH-71 presidential helicopter). This imposes some discipline, some of it applied in a laudable and hard-nosed manner, on a process that had run on autopilot for decades.

However, exemplary as some of these decisions may be, the surface has scarcely been scratched. One needs only to scan down the list of unaffected major defense acquisition programs currently in various stages of development or production to see, with few exceptions, a continuation of many horror stories similar to those that have plagued defense acquisition for decades. Not even all of the low hanging fruit has been removed.

What Is Needed?

There isn't much that knowledgeable observers of, and participants in, this process haven't already identified as problems and have proposed solutions for. They all appear in existing acquisition directives and instructions. Implementing them, rather than exercising their loopholes, is the starting point for fixing the process.[6]

With the current national fiscal environment and the lack of significant threats projected for the foreseeable future, waivers of the procedures and criteria for success that the regulations were designed to uphold should be few and far between, if they occur at all. In addition, they should be escalated to the

---

[6] Fundamentally, the directives and instructions specify three basic milestones with benchmarks required for approval to proceed into the next phase of the program: **Milestone A** – a decision to move into the technology development and demonstration phase, where system and sub-system prototypes are built and tested—also known as demonstration/validation (Dem/Val); **Milestone B** – formal program initiation with decision to proceed into Engineering and Manufacturing Development (EMD), previously called Full-Scale Engineering Development (FSED) or System Development and Demonstration (SDD); **Milestone C** – a production and deployment decision, starting with low-rate-initial production (LRIP) intended to provide production-representative systems for initial operational testing to support subsequent decisions to proceed with full-rate production (FRP) and deployment for initial operational capability (IOC).

Secretary of Defense for major, and even some lesser, programs. Finally, the Defense Department should not proceed with any program with waived requirements until the Congress and its independent arm, the GAO, have evaluated the rationale for the requested waivers, and the appropriate Congressional committees give explicit, statutory approval to proceed.[7] There is no rationale for not taking the necessary time for scrupulous analyses to determine whether we should embark on a new program. The responsibility and accountability must be clearly established and accepted at the top of the system.

The Front End: Setting Requirements

Hard-nosed discipline on the part of decision-makers at the front end of the process is crucial to reining in the appetite of the requirements community and precluding ill-informed development decisions based on immature technologies and optimistic projections of system costs, schedule and performance. Upfront realistic cost estimates and technical risk assessments, developed by independent organizations outside the chain of command for major programs, should inform Defense Acquisition Executives. The requirement for those assessments to be independent, not performed by organizations already controlled by the existing self-interested sections of the bureaucracy – as is the case now, even after WSARA 2009 – is essential.

The existing process has heartily approved presumed quantum leaps in claimed capability that are reflected in high-risk, often unattainable, technical and operational requirements. Many of these system performance goals have resulted from the salesmanship of the DOD research and development communities, combined with industry lobbying, in successfully convincing the user and the rest of the acquisition community that the hypothetical advanced capabilities could be delivered rapidly and cheaply.

In case after case, Pentagon decision-makers have acquiesced to programs entering FSED/EMD and even low-rate initial production before technical problems are identified, much less solved; before credible independent cost assessments are made and included in program budget projections; and before the more risky requirements are demonstrated in testing. This is nothing more than a "buy-in" to "get the camel's nose under the tent."

The MV-22 is a good example of a major program that encountered technical and cost problems after entering EMD in 1986, yet was approved to enter low-

---

[7] An accelerated version of this process can easily be designed to permit development and production for systems for the war in Afghanistan, but unjustified exploitation of the defense community's concern for the welfare of the troops must be prevented, and even the new accelerated process must include constant, independent oversight.

rate initial production (LRIP). In 1999, the declared urgency of replacing aging CH-46s drove decisions to severely reduce development testing before its completion, to enter operational testing prematurely and to gain approval for LRIP.

In April 2000, an MV-22 crashed during an operational test resulting in the deaths of 19 Marines. The official investigation into this tragic accident reported that the Flight Control System Development and Flying Qualities Demonstration (FCSDFQD) Test Plan investigating the phenomenon known as power settling was reduced from 103 test flight conditions to 49, of which only 33 were actually flight-tested with particularly critical test points not flown at all.

This series of events, culminating in the April 2000 accident and another crash in December of that year, brought the program to halt, nearly resulting in termination. However despite these setbacks, the program continued in low-rate production while Pentagon leadership debated whether to continue the program. In the end, the MV-22 program recovered, executed the full range of technical testing that should have been done previously, and was introduced into Marine Corps medium-lift forces in 2005, nearly 25 years after the decision to initiate the program. In the meantime, some 70 or more MV-22s had been procured, many of which required expensive modifications to correct deficiencies discovered in testing.

Among the Many False Reforms

The process has become even more cumbersome with the increased involvement of the Joint Chiefs of Staff (JCS). Over the years, the Joint Requirements Oversight Council (JROC) and the Joint Capabilities Integration and Development System (JCIDS) process were established to ostensibly provide the combat forces a greater voice in setting requirements. There is, however, little evidence that the "reformed" process has made any significant changes to programs as originally proposed by the advocates.[8]

---

[8] A report on January 2006, known as the Defense Acquisition Performance Report (DAPA) at https://acc.dau.mil/CommunityBrowser.aspx?id=18554 highlighted these continuing problems after decades of reform. Headed by retired Air Force Lt. Gen. Ronald Kadish, the panel found that "...the current requirements process does not meet the needs of the current security environment or the standards of a successful acquisition process. Requirements take too long to develop, are derived from Joint Staff and Services views of the Combatant Commands' needs and often rest on immature technologies and overly optimistic estimates of future resource needs and availability."

Real Reform: Considering Alternatives

Approval to proceed with any new development should depend on requirements, both technical and operational, that are attainable, affordable and testable and are based on realistic threat and funding projections. Most crucial to an effective new start is the conduct of an independent Analysis of Alternatives (AOA) that explores other approaches to meeting an identified need. The proposed solutions should run the gamut from continuing existing systems, to incremental improvements to those systems, to launching the development and procurement of a new system. DOD's regulations in Instruction 5000.2 call for AOAs to be completed and/or updated before each "Milestone" review, but in reality they have been few.[9]

A thorough AOA should be a hard and fast prerequisite for any milestone review. It should focus on an independent lifecycle cost estimate (R&D, procurement, and operating and support) and on the affordability of the various alternatives. It should also include realistic projections into the out years for cost, force levels, manpower support requirements, total procurement quantities, and affordable annual procurement rates. Done properly, an AOA should generate cost and schedule thresholds as well as key performance parameters (including measures of effectiveness, survivability, interoperability, and reliability and maintainability thresholds) upon which the rationale for a new program is based and where it fails in comparison to others. The performance thresholds should include both technical and operational measures that, in turn, should guide the planning and execution of both development and operational testing focused on those key parameters that constitute the justification for proceeding with the new program.

These independent analyses should be updated at regular intervals, not just for each program milestone. The process should be one of continuous evaluation, incorporating updated cost estimates and system performance projections, based on experience in development and testing to-date.

Periodic program assessments should weed out programs that are falling behind schedules, growing in cost and falling short of key measures of effectiveness and suitability.

Real Reform: Fly-Before-Buy/Competitive Prototype Demonstration

The "Fly-before-Buy" philosophy should be the mandated standard for all programs. Perhaps a better term would be "Fly-before-Decide." Done properly, it will demand the demonstration, through actual field testing of new

---

[9] WSARA 2009 also recognized this by calling for analyses that considered tradeoffs of cost, schedule and performance as part of the process for developing requirements.

technologies, subsystems, concepts, etc. to certain success criteria before proceeding at each milestone, not just the production decision. Accordingly, successful and competitive prototype development and testing should be a hard and fast prerequisite for any new development proceeding into the FSED/EMD phase. The Achilles heel of many defense programs has been their failure to adhere to this strategy, resulting in technical difficulties and costly development delays that could have been avoided had the decision-makers demanded successful completion of realistic prototype testing and evaluation.

Critical to the success of such a strategy is allocating sufficient up-front funding and schedule to permit a robust comparative evaluation of prototype systems in an operational environment during the Demonstration/Validation (Dem/Val) phase. The Defense Department has paid only lip service in the past to the competitive prototype demonstration requirement spelled out in its own directives. DOD should establish, adequately fund, and maintain operational units (e.g., aircraft squadrons, ground force brigades/battalions), independent of R&D organizations, to conduct tests and experiments to effect this concept.[10]

---

[10] Directly related to the "fly-before-buy" strategy are independent assessments of technology maturity or readiness levels before entering each stage of program development. It is crucial to any successful development program that appropriate levels of technology maturity/readiness be demonstrated, primarily through testing of systems and subsystems (as opposed to paper studies or simulations), before decisions to proceed to a given stage in program development. The July 2009 DOD Technology Readiness Assessment (TRA) Deskbook (at http://www.dod.mil/ddre/doc/DoD_TRA_July_2009_Read_Version.pdf) spells it out. The purpose is to provide the decision-maker with an independent view of whether or not the technologies embodied in a new system have demonstrated appropriate levels of maturity to justify proceeding into the next phase of development or procurement. The Deskbook provides definitions of the nine technology readiness levels (TRLs) to be used in independent evaluations of critical technology maturity. The Deskbook spells out specific TRLs to be demonstrated for the critical program milestones B and C. Milestone B approval, or entry into EMD, requires TRL level 6 to include a "representative model or prototype system … is tested in a relevant environment. Examples include testing a prototype in a high fidelity laboratory environment or in a simulated operational environment." Unfortunately, this criterion does not go far enough. Rather, the process should be altered to demand demonstration of TRL 7, defined in the Deskbook as " Prototype near or at planned operational system … requiring the demonstration of an actual system prototype in an operational environment, such as an aircraft, vehicle, or space."
In a similar vein, TRL 7, required for successful entry into Low-Rate Initial Production (LRIP) at Milestone C, is insufficient: " Prototype near or at planned operational system" does not go far enough in ensuring the readiness of a system for production. Rather, the success criterion for LRIP approval should depend on an independent assessment that TRL 8 has been achieved: "Technology has been proven to work in its final form and under expected conditions. In almost all cases, this TRL represents the end of true system development. Examples include development test and evaluation of the system in its intended weapon system, to determine if it meets design specifications." Without

Real Reform: Event-Based, Not Schedule-Based Decisions

DOD's experience with systems entering operational testing prior to completion of sufficient development testing is chronicled in innumerable GAO and several Defense Science Board (DSB) reports in recent years. A May 2008 DSB Task Force Report on Development Test and Evaluation found that, in the ten year period between 1997 and 2006, over two-thirds of Army systems failed to meet their reliability requirements in operational testing.[11] In almost all these cases, the systems had entered operational test and evaluation (OT&E) with little or no chance of success, based on the failures demonstrated during development testing. These programs had not met the criteria for successful completion of development testing and had entered OT&E doomed to fail.

The acquisition decision authority should impose an event-based strategy on programs with meaningful and realistic pass/fail criteria for each stage of development and production. Only if the criteria are satisfied (through actual testing where applicable) should the decision-maker allow a program to proceed to its next phase. For example, when a program is approved at Milestone B to move into EMD, approval to successfully pass a future Milestone C and proceed into low-rate initial production should be predicated on the program demonstrating specific performance/reliability/cost thresholds.
Failure to achieve these goals should result in program termination or at least significant restructure until they are met.

Real Reform: Continuous Evaluations

As a new program begins, a process of continuous and independent evaluation must be established to track the program through development, testing and production, and eventual fielding and employment by operational forces. In the early stages, such evaluations should be based on emerging test results and updated cost estimates, and should focus on those attributes or capability measures that formed the basis for program approval. These evaluations should be updated with results presented to senior leadership on a routine basis—certainly at least annually. Such evaluations should inform decisions whether or

---

question, a new system should not be put into production until development testing has shown that the design is complete and proven to work.
As currently implemented, the evaluations of technology maturity and assignment of TRLs are the responsibility of the Research and Technology organization in the Pentagon, with input from the test community. This arrangement casts doubt on the true independence of the TRAs. A more appropriate approach would have the testing community tasked with final responsibility for the independent TRAs at Milestones B and C.

[11] See the May 2008 Defense Science Board Task Force on Development Testing and Evaluation at http://www.acq.osd.mil/dsb/reports/ADA482504.pdf.

not to proceed with the program or to restructure the program goals and acquisition strategy.

It is extremely important that this process of continuous evaluation extend beyond development. Organizations, independent of both the development and operational communities, should be established and maintained to track experience with new and existing systems in the field, evaluating data gathered in training sorties and exercises as well as in combat, where applicable. Assessments should include not only the usual measures of system performance, but also all aspects of system supportability, to include reliability, availability and maintainability (RAM), as well as safety, training and human factors.

Feedback loops from the field to the requirements and acquisition communities should be established and maintained throughout the life of a weapon or system. Such arrangements should take full advantage of operational experience in developing plans and requirements for starting a new program, determining needed fixes for deficiencies encountered in the field, and continuing and/or upgrading existing systems. Such lessons learned should be invaluable to the acquisition community in shaping its approach to the development of new systems as well as to the test and evaluation and analytic communities in structuring their evaluations of similar systems in the future.

Conclusion

As the country enters what promises to be a prolonged period of fiscal austerity, it can no longer afford the extravagance of spending hundreds of billions of dollars and not receiving the capabilities it paid for. Fortunately, we have an extensive base of experience, derived from both military and commercial programs that we can draw upon to avoid the mistakes of the past. These lessons have been codified in DOD regulations, and the evidence shows that the vast majority of cost overruns and schedule delays come from avoiding their requirements, particularly in the initial stages of a program.

We are also fortunate that there is no need to rush new systems into development and procurement in order to counter some imminent new threat. The F-16, for example, entered operational service in 1980 and is still in production. It and the remaining A-10s in the Air Force's inventory are more than adequate aircraft for existing missions in Afghanistan and for conventional threats, should they arise. There is no projected threat on the horizon that would justify taking additional risk by compressing development schedules for any new system (such as the highly problematic F-35 program). Moreover, compressing prescribed schedules when real threats actually exist, such as during the Cold War, has proven to be a huge cost and performance disaster – and to save no time.

We have the tools and expertise we need to make substantial reductions in the cost overruns, performance disappointments and schedule slips that plague our weapon programs. What we do not have, or have not had consistently, is the determination to apply the available tools, especially when it means canceling programs that are generating careers in the Pentagon and jobs, campaign contributions and votes outside it.

Suggested Contacts, Readings and Web Sites

Contacts

Contact the various authors of this handbook at the following e-mail addresses:

Thomas Christie: tchristie34@verizon.net
Andrew Cockburn: amcockburn@gmail.com
Bruce Gudmundsson: trossknecht@yahoo.com
Chet Richards: FuentesDeOnoro@me.com
Franklin C. Spinney: chuck.spinney@gmail.com
Pierre M. Sprey: Pierre@mapleshaderecords.com
Winslow Wheeler: winslowwheeler@msn.com
George Wilson: gcwilson1@comcast.net
G.I. Wilson: wilsongi@aol.com

Readings

Each of the authors was asked to recommend readings; what follows is our compilation. Annotations in quotes are excerpts from various materials at Amazon.com. Comments by various authors are identified as such or simply lack quotation marks.

Human Conflict

Robert Coram, *Boyd: The Fighter Pilot Who Changed the Art of War* (Little, Brown and Company, 2002). "John Boyd (1927-1997) was a brilliant and blazingly eccentric person. He was a crackerjack jet fighter pilot, a visionary scholar and an innovative military strategist. Among other things, Boyd wrote the first manual on jet aerial combat, was primarily responsible for designing the F-15 and the F-16 jet fighters, was a leading voice in the post-Vietnam War military reform movement and shaped the smashingly successful U.S. military strategy in the Persian Gulf War. His writings and theories on military strategy remain influential today, particularly his concept of the 'OODA (Observation, Orientation, Decision, Action) Loop,' which all the military services-and many business strategists-use to this day. Boyd also was a brash, combative, iconoclastic man, not above insulting his superiors at the Pentagon (both military and civilian); he made enemies (and fiercely loyal acolytes) everywhere he went...."

For a concise summary of Boyd's work, see the entry below.

Franklin C. Spinney, "Genghis John," *Proceedings*, U.S. Naval Institute, July 1997, 42-47, http://pogoarchives.org/labyrinth/01/01.pdf. Almost no one understood John Boyd better than "Chuck" Spinney; this article concisely describes Boyd's life, thinking and legacy.

Grant T. Hammond, *The Mind of War: John Boyd and American Security* (Smithsonian Institution Press, 2001). "Breakthrough biography of a revolutionary thinker who transformed American military policy and practice. Based on extensive interviews with Boyd and with those who knew him, *The Mind of War* is the first biography of this pivotal figure in American military history."

Col. Chet Richards, *Certain to Win: The Strategy of John Boyd Applied to Business* (Center for Defense Information, 2004). "Develops the strategy of the late US Air Force Colonel John R. Boyd for the world of business. Robert Coram's monumental biography, *Boyd, the Fighter Pilot Who Changed the Art of War*, rekindled interest in this obscure pilot and documented his influence on military matters ranging from the design of the F-15 and F-16 fighters to the planning for Operation Desert Storm…."

Find the major elements of Boyd's work at http://dnipogo.org/john-r-boyd/, including Boyd's *Discourse on Winning and Losing,* which includes his *Patterns of Conflict, Strategic Game of ? and ?, Organic Design for Command and Control*, and *The Essence of Winning and Losing*. Boyd's methods, as he explained them in *Discourse* and other materials, enable people—from a single individual to an alliance of nations—to orient themselves to external challenges and opportunities, create options, take actions and exploit their effects before their opponents can understand and react effectively. In order to do this, Boyd explained, they employ certain active and passive measures to keep their common implicit orientation better harmonized both among themselves and with external reality than their opponents.

Find various YouTube videos of sections of John Boyd briefings and materials about him (of varying quality) at
http://www.youtube.com/results?search_query=John+Boyd&aq=f.

Sun Tzu, *The Art of War*, ed. and trans. John Minford (Penguin Books, 2002). "The Art of War is among the greatest classics of military literature ever written. Sun Tzu warfare is as applicable today as when the book was written some 2,500 years ago...Pick up The Art of War and read it." – Gen. A.M. Gray, former U.S. Marine Corps commandant, *Marine Corps Gazette*.

Carl von Clausewitz, *On War* (Multiple publications and publishers). "Written two centuries ago by a Prussian military thinker, this is the most frequently cited, the most controversial, and in many ways, the most modern book on warfare. The author fought against the armies of the French Revolution and Napoleon, served as a staff officer, and became a prominent military educator. In this work, he examines moral and psychological aspects of warfare, stressing the necessity of courage, audacity and self-sacrifice, as well as the importance of morale and public opinion. He emphasizes the notion of strategy as an evolving plan, rather than a formula, a concept adaptable to modern strategists in fields beyond military science."

People

Robert D. Hare and Paul Babiak, *Snakes In Suits; When Psychopaths Go to Work* (Collins Business, 2006). "Psychopaths are described as incapable of empathy, guilt, or loyalty to anyone but themselves; still, spotting a psychopath isn't easy…A common description of psychopathology states that subjects 'know the words but not the music;' Babiak and Hare state that 'a clever psychopath can present such a well-rounded picture of a perfect job candidate that even seasoned interviewers' can be fooled…to illuminate the power of the psychopath to manipulate those around him, as well as what strategies can be used to identify and disarm him."

Irving L. Janis, *Groupthink: Psychological Studies of Policy Decisions and Fiascoes* (Houghton Mifflin, 1983). "Janis defines groupthink as the 'deterioration of mental efficiency, reality testing, and moral judgment' in the interest of group solidarity. Pressure to conform…Group members tend to show strong favoritism toward their own ideas in the manner by which information is processed and evaluated, thus guaranteeing that their ideas will win out."

Jonathan Shay, *Achilles in Vietnam: Combat Trauma and the Undoing of Character* (Scribner, 1985). "War…generates rage because of its intrinsic unfairness. Only one's special comrades can be trusted. The death of Patroklos drove Achilles first into passionate grief, then into berserk wrath. Shay establishes convincing parallels to combat in Vietnam, where the war was considered meaningless and mourning for dead friends was thwarted by an indifferent command structure...recommends policies of unit rotation and unit "griefwork"—official recognition of combat losses—as keys to sustaining…a moral existence during war's human encounters. The alternatives are unrestrained revenge-driven behavior, endless reliving of the guilt such behavior causes and the ruin of good character."

Jonathan Shay, *Odysseus in America: Combat Trauma and the Trials of Homecoming* (Scribner, 2003). "…uses Odysseus's epic journey to explore the stresses faced by veterans who return home, still scarred by their intense experiences…Odysseus experienced nearly all of the symptoms he has observed in returned veterans of modern wars: fearfulness, inability to trust or be close to anyone, emotional outbursts, violence, criminal activity, sexual adventurism, and so forth…deals with healing techniques...[and] suggested measures for prevention of such long-lasting injuries…"

Martin van Creveld, *Fighting Power: German and U.S. Army Performance, 1939–1945* (Greenwood Press, 1982). "…analyzes the ways in which the WWII German Army developed the fighting power that allowed them to achieve a number of military victories even when outnumbered and using outdated equipment. He compares and contrasts the Germans with the U.S. Army, which developed a different style of war based on superior economic and technological resources."

Maj. Donald Vandergriff, U.S. Army, ret., *The Path to Victory: America's Army and the Revolution in Human Affairs* (Presidio Press, 2002). "Instead of just analyzing the problem [in the officers' corps], Vandergriff gives us the foundation for a new system."

Maj. Donald Vandergriff, U.S. Army, ret., *Raising the Bar: Creating and Nurturing Adaptability to Deal with the Changing Face of War* (Center for Defense Information, 2006). "'Adaptability' has become a buzzword throughout the U.S. Army due to experiences in Afghanistan and Iraq…The Army recognizes that in order to move toward becoming a 'learning organization' where leaders practice adaptability, it will have to change its culture, particularly its leader development paradigm."

Maj. Donald Vandergriff, U.S. Army, ret., *Manning the Future Legions of the United States: Finding and Developing Tomorrow's Centurions* (Praeger, 2008). "…looks beyond recruiting. It is a holistic view of today's Army and addresses the fact that in order to effectively recruit the soldiers and leaders of the future, the nation needs to take the Army—its personnel management system and structure—from the Industrial Age into the Information Age."

George C. Wilson, *Mud Soldiers: Life Inside the New American Army* (Collier Books, 1991). "Wilson became increasingly critical of the Army as he accompanied a group of volunteers through basic and advanced infantry training and their first field maneuver; mishandled trainees, improper protection during exercises and four suicide attempts led him to recommend program changes."

The Pentagon and Military Reform

James G. Burton, *The Pentagon Wars: Reformers Challenge the Old Guard* (U.S. Naval Institute Press, 1993). "…testifies that the process of selecting and purchasing weapons for our armed forces is 'ethically and morally corrupt from top to bottom,' with few checks and balances. The most scathing and damning portions of the expose illustrate how Pentagon procurement officers routinely give more consideration to satisfying defense contractors than to the safety of the troops who will use a given weapon on the field..." Also contains an epilogue explaining how the Republican Guard was allowed to escape at the end of the First Gulf War, thereby enabling Saddam Hussein's regime to survive—necessitating the Second Gulf War.

Thomas Christie, "What Has 35 Years of Acquisition Reform Accomplished?" *Proceedings*, U.S. Naval Institute, February 2006, http://www.cdi.org/pdfs/Christie%20in%20Proceedings.pdf. Based on 40 years of experience in the system, Christie explains why the Defense Department's acquisition system has become such a pervasive failure and why persistent changes have failed.

Andrew Cockburn, *Rumsfeld: His Rise, Fall, and Catastrophic Legacy* (Scribner, 2007). "Relying on sources that include high-ranking officials in the Pentagon and the White House, *Rumsfeld* goes far beyond previous accounts to reveal a man consumed with the urge to dominate each and every human encounter, and whose aggressive ambition has long been matched by his inability to display genuine leadership or accept responsibility for egregious error...Cockburn reveals how Rumsfeld's habits of intimidation, indecision, ignoring awkward realities, destructive micromanagement and bureaucratic manipulation all helped doom America's military adventure."

James Fallows, *National Defense* (Random House, 1981). This 30-year-old classic introduces many of the individuals, concepts and techniques of military reform.

Ernest Fitzgerald, *The High Priests of Waste* (Norton, 1972).; Ernest Fitzgerald, *The Pentagonists: An Insider's View of Waste, Mismanagement and Fraud in Defense Spending* (Houghton Mifflin, 1989). These two volumes are "Ernie" Fitzgerald's descriptions of the obstacles he met when trying to expose waste and fraud in the Pentagon. His efforts earned him numerous efforts by his superiors, including Richard Nixon, to fire him—all of them unsuccessful.

William S. Lind and Gary Hart, *America Can Win: The Case for Military Reform* (Adler & Adler, 1986). Lind and Hart present the basic case for military reform.

Col. Douglas Macgregor, U.S. Army, ret., *Transformation Under Fire: Revolutionizing How America Fights* (Praeger, 2003). "Macgregor's book is in the best tradition of military theorists, whose ideas transformed armies to meet the challenges of WWII: Hans Von Seckt, B. H. Liddell Hart, Charles de Gaulle, and Heinz Guderian. Macgregor presents the first coherent view of how the information age should transform the way we organize for war…takes to task the leadership culture that stifles change..." – *ARMOR Magazine.*

Andrew Pasztor, *When the Pentagon Was for Sale: Inside America's Biggest Defense Scandal* (Scribner, 1995). "Pasztor's examination of Pentagon and arms-industry corruption exposes the process by which such giant defense contractors as Boeing, General Electric and United Technologies illegally obtained contracts with the cooperation of Pentagon officials throughout the Reagan years...tracks the criminal investigations and prosecution of defense suppliers and Pentagon officials during the Justice Department's Operation Illwind...he maintains that very little has changed to improve day-to-day accountability, and the Pentagon's own rules and regulations continue implicitly to encourage wrongdoing."

Dina Rasor, *More Bucks Less Bang: How the Pentagon Buys Ineffective Weapons* (Bookpeople, 1983).; Dina Rasor, *Pentagon Underground* (Crown, 1985). With help from Pentagon insiders, Rasor uncovered horror story after horror story about prodigious Pentagon waste and inept, and sometimes corrupt, management.

Col. Chet Richards, *A Swift Elusive Sword: What if Sun Tzu and John Boyd Did a National Defense Review?* (Center for Defense Information, 2001). "…suggests that ancient strategic wisdom may help solve the dilemma confronting the U.S. military: spending on defense exceeds that of any combination of potential adversaries, but the services still face cancellation of weapon systems and lack of funds for training, spares, and care and feeding of the troops. Richards suggests U.S. military leaders can break out of the 'dollars equals defense' mindset, and create more effective forces…"

Franklin C. Spinney, *Defense Facts of Life: The Plans/Reality Mismatch* (Westview Press, 1985). "Well-documented and well-illustrated account of how virtually every single weapons and mobility system now in the Pentagon system is over-priced, over-weight, over-budget, and not able to perform as advertised…[T]he author is very effectively demonstrating that doctrine, technology and the budget are completely divorced from both real world threats, and real world logistics…"

Some of the other works by "Chuck" Spinney are listed below. These briefings, essays and articles (and Spinney's book listed above) contemporaneously document the affirmation that today's defense problems are no accident and could have been, indeed were, foreseen. Moreover, all of the problems could have been mitigated or avoided by senior management in the Pentagon or by Congress if either had the character to orient to what was best for the country rather than what was best for political and bureaucratic careers, and membership in good standing in elite decision-making circles.

Franklin C. Spinney, "Genghis John," *Proceedings*, U.S. Naval Institute, July 1997, 42-47, http://pogoarchives.org/labyrinth/01/01.pdf.

Franklin C. Spinney, Statement before the Subcommittee on National Security, Veterans Affairs and International Relations, Committee on Government Reform, U.S. House of Representatives, June 4, 2002, http://pogoarchives.org/labyrinth/01/02.pdf.

Franklin C. Spinney, "The New QDR: The Pentagon Goes Intellectually AWOL," *CounterPunch*, February 2010, http://pogoarchives.org/labyrinth/01/03.pdf.

Franklin C. Spinney, "The JSF: One More Card in the House," *Proceedings of the U.S. Naval Institute*, August 2000, http://pogoarchives.org/labyrinth/01/04.pdf.

Franklin C. Spinney, "Defense Death Spiral," September 1998, http://pogoarchives.org/labyrinth/01/05.pdf.

Franklin C. Spinney, "Porkbarrels & Budgeteers: What Went Wrong with the Defense Review," September 1997, http://pogoarchives.org/labyrinth/01/06.pdf.

Franklin C. Spinney, "Defense Time Bomb; Background: F-22/JSF Case Study Hypothetical Escape Option," March 1996, http://pogoarchives.org/labyrinth/01/07.pdf.

Franklin C. Spinney, "Three Reasons Why the ATF Should Not Be Approved for Engineering and Manufacturing Development," July 23, 1991, http://pogoarchives.org/labyrinth/01/08.pdf.

Franklin C. Spinney, "Defense Power Games," October 1990, http://pogoarchives.org/labyrinth/01/09.pdf. .

Maj. Donald Vandergriff, U.S. Army, ret., *Spirit, Blood and Treasure* (Presidio Press, 2001). "The new millennium brings with it a need for unprecedented flexibility and responsiveness in our national defense apparatus. In the view of this expert panel we are nowhere near ready."

Winslow T. Wheeler et al., *America's Defense Meltdown: Pentagon Reform for President Obama and the New Congress* (Stanford University Press, 2009).

"...describes how America's armed forces are manned and equipped to fight, at best, enemies that do not now—and may never again—exist and to combat real enemies ineffectively at high human and material cost...Over time, policy makers of all political stripes have created budgets that have made our forces smaller, less well equipped, and less ready to fight—all at dramatically increasing cost. Fortunately, the book's authors offer 'real-world' solutions to all the problems they identify..." Stanford University Press has permitted three chapters of the anthology to be downloadable at the *Labyrinth* websites: find Col. Chet Richards' chapter on strategy at http://pogoarchives.org/labyrinth/11/01.pdf; Col. G.I. Wilson's and Maj. Don Vandergriff's chapter on people issues at http://pogoarchives.org/labyrinth/11/02.pdf, and Col. Bob Dilger's and Pierre Sprey's chapter on air power at http://pogoarchives.org/labyrinth/11/03.pdf.

Winslow T. Wheeler and Lawrence J. Korb, *Military Reform: An Uneven History and an Uncertain Future* (Stanford Security Studies, 2009). "establish[es] a definition of what genuine military reform is and is not, and [identifies] what *really* needs to be done to transform our military. They compare genuine reform with 'cosmetic dabbling'—that improves nothing and often burdens US combat forces to the point of mental and physical immobility..."

George C. Wilson, *This War Really Matters: Inside the Fight for Defense Dollars* (Congressional Quarterly Press, 2000). "Drawing on nearly 40 years of news writing focused on military issues, George C. Wilson takes the reader through a fascinating, but little understood, process: how the Pentagon and Congress spend $500,000 a minute on guns and soldiers. Interweaving personal stories and insights from the major players..., Wilson provides an inside look at how the 105th Congress and the Pentagon battled for a 250 billion dollar defense budget."

Conventional and Maneuver Warfare

Larry H. Addington, *Patterns of War through the Eighteenth Century* (Indiana University Press, 1990).; Larry H. Addington, *Patterns of War since the Eighteenth Century* (Indiana University Press, 1994). Bruce Gudmundsson describes both as an excellent overview of the evolution of the Western way of war. The second volume, which deals with the past two centuries, is more detailed, and provides an excellent companion to *The American Way of War* (see below).

Carl H. Builder, *The Masks of War: American Military Styles in Strategy and Analysis* (The Johns Hopkins University Press, 1989). Bruce Gudmundsson describes this as the single best description of the way that the various armed services look at the world—an indispensable tool for anyone who finds himself working with any of the armed services.

James F. Dunnigan and Austin Bay, *A Quick and Dirty Guide to War: Briefings on Present and Potential Wars* (Paladin Press, 2008). Bruce Gudmundsson describes this as an excellent introduction to the wars of the past 30 years or so. True to its title, this book is a lively work aimed at a broad audience, and is thus well suited to someone new to the subject.

Trevor N. Dupuy, *A Genius for War: The German Army and General Staff, 1807–1945* (Military Book Club, 1977). There are many books that competently address the organizational concepts and the "styles of warfare" that the Germans adopted beginning in the 19th century and that were refined late in World War I and shortly thereafter—thereby laying the basis for the extraordinarily successful "blitzkrieg" form of warfare that Germany's World War II opponents had to attempt to emulate and adapt to in order to compete. Dupuy's book listed here is one of several that are available.

Bruce I. Gudmundsson, *Storm Troop Tactics: Innovation in the German Army, 1914-1918* (1989).; Bruce I. Gudmundsson, *On Artillery* (Praeger, 1993).; Bruce I. Gudmundsson and John A. English, *On Infantry* (Praeger, 1995).; and Bruce I. Gudmundsson, *On Armor* (Praeger, 2006). Intended for military professionals, but quite readable for the student, each of these books addresses the roots of successful mental and material innovation in modern military forces.

Ernst Junger, *Storm of Steel.* (Penguin. 2002). A World War I memoir by a young German officer who, survived the front line on the western front all the way through. It is a searing depiction of modern war, and it is an indispensable source on trench warfare and the tactics evolved by the Germans in dealing with it.

William S. Lind, *Maneuver Warfare Handbook* (Westview Special Studies in Military Affairs, 1985). "Maneuver warfare, often controversial and requiring operational and tactical innovation, poses perhaps the most important doctrinal questions currently facing the conventional military forces of the U.S. Its purpose is to defeat the enemy by disrupting the opponent's ability to react, rather than by physical destruction of forces…The authors translate concepts too often vaguely stated by maneuver warfare advocates into concrete doctrine."

Col. Douglas Macgregor, U.S. Army, ret., *Warrior's Rage: The Great Tank Battle of 73 Easting* (Naval Institute Press, 2009).; and Col. Douglas Macgregor, U.S. Army, ret., *Breaking the Phalanx: A New Design for Landpower in the Twenty-first Century* (Praeger, 1987). *Warrior's Rage* "…recounts two stories. One is the inspiring tale of the valiant American soldiers, sergeants, lieutenants, and captains who fought and won the battle. The other is a story of failed generalship, one that explains why Iraq's Republican Guard escaped, ensuring that Saddam Hussein's regime survived and America's war with Iraq dragged on. Certain to provoke debate, this is the latest book from the controversial and influential military veteran whose two previous books, *Breaking the Phalanx* and *Transformation Under Fire*, are credited with influencing thinking and organization inside America's ground forces and figure prominently in current discussions about military strategy and defense policies."

Bruce Porter, *War and the Rise of the State* (Free Press, 2002). Bruce Gudmundsson describes it as laying out the interrelationship between the rise of the state as an institution and the way that wars have been fought in the past five centuries. In doing so, it makes the classic argument that the modern state and modern armies grew in tandem and that, in particular, the welfare state is largely a product of the total wars of the 20th century.

Russell F. Weigley, *The American Way of War: A History of United States Military Strategy and Policy* (Indiana University Press, 1977). The book argues that there are two distinct traditions in the way the United States fights wars, one based upon maneuver and the other upon the massive application of firepower. In the course of doing this, it also provides an accessible introduction to American military history.

Insurgency and Fourth Generation Warfare

Robert B. Asprey, *War in the Shadows: The Guerrilla in History* (Doubleday, 1975). "… survey of guerrilla warfare begins with the struggle between Persian king Darius and Scythian irregulars and concludes with the mujahedin resistance to the Soviet invasion of Afghanistan. He discusses how great commanders such as Hannibal and Napoleon dealt with irregulars and how counterinsurgency experts such as Sir Gerald Templar during the Malayan Emergency in the early 1950s found ways to defeat the guerrilla."

Thomas X. Hammes, *The Sling and the Stone* (Zenith Press, 2004). A classic, this is one of several contemporary quality analyses available on insurgent warfare.

T.E. Lawrence, *Seven Pillars of Wisdom: A Triumph* (Anchor, 1991). This is another classic description of insurgent warfare. Advocates of "Fourth Generation War" as something entirely new are well advised to review T. E. Lawrence's descriptions of it in the First World War.

William S. Lind et al., "The Changing Face of War: Into the Fourth Generation," *The Marines Corps Gazette*, October 1989. This 1989 article includes a prediction of the emergence—more than 10 years later—of what has become known as "Fourth Generation War."

William R. Polk, *Violent Politics: A History of Insurgency, Terrorism, and Guerrilla War, from the American Revolution to Iraq* (Harper, 2008). "...insurgencies throughout history, beginning with America's own struggle for independence...Polk explores the role of insurgency in other notable conflicts—including the Spanish guerrilla war against Napoleon, the Irish struggle for independence, the Algerian War of National Independence, and Vietnam—eventually landing at the ongoing campaigns in Afghanistan and Iraq, where the lessons of this history are needed more than ever."

John H. Poole, A series of books: *The Last Hundred Yards; One More bridge to Cross; Phantom Soldier; The Tiger's Way; Tactics of the Crescent Moon; Militant Tactics; Terrorist Trails; Dragon Days; Expeditionary Eagles; Homeland Siege; Tequila Junction.* "The entire Poole series, beginning with *The Last Hundred Yards* . . . through his most recent *Tequila Junction* . . . provide unique insight into terrorists, insurgents, and guerrillas that is underappreciated within defense and security hierarchies." –Counterterrorist Magazine, January 2009. "John Poole has written a thought-provoking and intriguing work in *Tequila Junction.* He has masterfully made the case for attention and action toward threats being ignored due to our myopic focus on Islamic extremism. This is another exceptional volume to add to his superb collection of works dealing with the new forms of conflict we face." – Gen. Anthony C. Zinni USMC (Ret.), June 2008

Col. Douglas Macgregor writes as follows on this series: Poole's 'series' involve light infantry tactics and techniques at the squad level and sometimes at the platoon level. Though his work at the squad and platoon level has value, it should stay at that level. Nothing of what he writes is applicable to modern warfare involving opponents with the capability to fight back.... Poole's principle mistake is in believing improvement at the light infantry squad or platoon level will lead American forces to victory in the so called 4GW and counterinsurgency.... Given that we now live in an age of extreme battlefield lethality, the approach Poole advocates is terribly archaic and dangerous. It promises to produce heavy casualties against any enemy with capability above small arms. History demonstrates repeatedly that light infantry has no chance against modern armored forces. We saw this in the Pacific with the Imperial Japanese Army where the pathetic Sherman tank and American air power devastated the Japanese forces. We also saw it during the initial failure of the US and ROK ground forces defending against tank units of the North Korean army....In the end firepower, mobility, and armored protection, the holy trinity of offensive military power in land warfare, is decisive. When augmented with effective strikes from the air, it is irresistible....In summary, Poole is selling dangerous snake oil and miracle cures that don't exist. Poole's approach may seem to promise savings in defense spending to those who just want to cut spending, but it's a guaranteed loser in a real war with real armies, real air forces, and real air defenses. Buyer beware!

Col. Chet Richards, U.S. Air Force, ret., *Neither Shall the Sword: Conflict in the Years Ahead* (Center for Defense Information, 2006). "Despite spending on defense that equals the rest of the world, combined, and initiating a war in Iraq that will likely surpass Vietnam in cost, the United States has yet either to destroy al-Qaeda or to defeat a group of ragtag insurgents concentrated in the areas around Baghdad. The U.S. Department of Defense…is not only unsuited for this new form of conflict, it cannot be transformed into an organization that is."

Martin van Creveld, *The Transformation of War* (The Free Press, 1991). "Most wars since 1945 have been low-intensity conflicts and, according to the author, incomparably more significant than conventional wars in terms of casualties suffered and political results achieved...Van Creveld, who teaches history at Hebrew University in Jerusalem, argues that the theories of Karl von Clausewitz, which form the basis for Western strategic thought, are largely irrelevant to nonpolitical wars such as the Islamic jihad and wars for existence such as Israel's Six-Day War...Weapons will become less, rather than more, sophisticated and the high-tech weapons industry (which 'supports itself by exporting its own uselessness') will collapse like a house of cards..."

Intelligence

James Bamford, *The Puzzle Palace: Inside the National Security Agency, America's Most Secret Intelligence Organization* (Penguin, 1983).; James Bamford, *The Shadow Factory: The NSA from 9/11 to the Eavesdropping on America* (Anchor, 2009). "James Bamford has been the preeminent expert on the National Security Agency since his reporting revealed the agency's existence in the 1980s. Now Bamford describes the transformation of the NSA since 9/11, as the agency increasingly turns its high-tech ears on the American public."

Andrew Cockburn, *The Threat: Inside the Soviet Military Machine* (Random House, 1983). Cockburn's analysis of how the U.S. intelligence community ignored readily available data to assess actual Soviet military capabilities—choosing instead to cooperate with politically directed "threat inflation"—pertains to the bygone era of the Cold War. However, the book remains most relevant today for understanding how contemporary threats can remain so poorly understood (and in many cases highly inflated beyond their actual capabilities).

David Kahn, *The Codebreakers: The Comprehensive History of Secret Communication from Ancient Times to the Internet* (Scribner, 1996). "Most of

The Codebreakers focuses on the 20th century, especially World War II. But its reach is long. Kahn traces cryptology's origins to the advent of writing."

Weapons and Technology

Thomas S. Amlie, "Radar: Shield or Target," *IEEE Spectrum*, April 1982. Amlie points out in this seminal article that what many think to be a key to effectiveness in military technology (radar) also has many limitations and negative trade-offs: Like a flashlight at night, the Aegis radar can see, but it can be seen, and tracked, from much further. Available at http://pogoarchives.org/labyrinth/11/10.pdf.

Government Accountability Office (GAO), *Operation Desert Storm: Evaluation of the Air Campaign*, June 1997, U.S. General Accounting Office, GA)/NSIAD-97-134. Not easy-to-read and poorly organized, the appendices of this 200-page analysis of the air campaign of the First Gulf War are crammed with Air Force and Navy data to assess the actual—not hyped—performance of high- and low-tech systems in that air war. Virtually all of the impressive claims in favor of extraordinary performance of "precision," high-tech systems—and especially of a "revolution in warfare" occurring—were not just unsupported by the facts but refuted.

Lt. Col. Patrick Higby, U.S. Air Force, *Promise and Reality: Beyond Visual Range (BVR) Air-To-Air Combat*, Air War College, Seminar 7, Maxwell Air Force Base, AL, http://www.vmi.edu/uploadedfiles/archives/adams_center/essaycontest/20042005/higbyp_0405.pdf. Also available at http://pogoarchives.org/labyrinth/11/09.pdf, it addresses the widely divergent theory and practice of beyond visual range radar-based air-to-air missiles and their consistent failure in air-to-air combat from Korea to Operation Desert Storm.

Charles E. Myers, "Air Support for Army Maneuver Forces," *Armed Forces Journal*, March 1987. "Chuck" Myers conducted seminars on what he and other military reformers consider to be the core mission of air forces in warfare to most directly impact the outcome of the war: "close air support," or direct support to ground forces in contact with the enemy, the classic mission of the German "Stuka" in World War II and of the U.S. Air Force A-10 in the wars in the Persian Gulf and Afghanistan.

Col. Everest E. Riccioni, U.S. Air Force, ret., "Strategic Bombing: Always a Myth," *Proceedings*, U.S. Naval Institute, November 1996. Available at http://pogoarchives.org/labyrinth/11/11.pdf. One of the original "fighter mafia" that started the reform movement, Colonel Riccioni, like John Boyd, was a brilliant and innovative aircraft designer. This article argues "Seventy-five years of praying at the altar of Giulio Douhet—the god of strategic bombing—has proved worthless. We must assess bombing theory and practice analytically, and develop a new model for the future."

Pierre M. Sprey, "The Case for More Effective, Less Expensive Weapon Systems: What 'Quality Versus Quantity' Issue?" *The Military Reform Debate: Directions for the Defense Establishment for the Remainder of the Century*, Background Pamphlet, United States Military Academy, West Point, NY, June 3–5, 1982 (available at http://pogoarchives.org/labyrinth/11/12.pdf); and Pierre M. Sprey, "Land-Based Tactical Aviation," *Critical Issues: Reforming the Military*, ed. Jeffrey G. Barlow, The Heritage Foundation, 1981 (available at http://pogoarchives.org/labyrinth/11/06.pdf). Much of Sprey's written work is unpublished but exists in the form of hard copies of briefing slides, with subjects address such as the so-called 'Quality vs. Quantity' Issue (now available at http://pogoarchives.org/labyrinth/11/08.pdf), and titles like "Combat Lessons from Lebanon and the Falklands: Is There a Little Wheat under All That Chaff?" (at http://pogoarchives.org/labyrinth/11/04.pdf) and "Letting Combat Results Shape the Next Air-to-Air Missile" (at http://pogoarchives.org/labyrinth/11/07.pdf). Additional titles by Pierre Sprey are available at the documents listed for Essay #9 of this handbook available at http://dnipogo.org/labyrinth/.

James Stevenson, *The Pentagon Paradox: The Development of the F-18 Hornet* (U.S. Naval Institute Press, 1993).; James Stevenson, *The $5 Billion Misunderstanding: The Collapse of the Navy's A-12 Stealth Bomber Program* (U.S. Naval Institute Press, 2001). Stevenson's two comprehensive treatises address much more than just the history of the aircraft in the titles. He also addresses the concepts behind successful fighter designs and how the aviation bureaucracy in the Pentagon is willing to embrace, instead, poorly conceived fighter and bomber aircraft and then spend untold billions of dollars to bring those bureaucratically and politically driven aircraft concepts to fruition.

Martin van Creveld, *Technology and War: From 2000 BC to the Present* (The Free Press, 1989). "Van Creveld considers man's use of technology over the past 4,000 years and its impact on military organization, weaponry, logistics, intelligence, communications, transportation, and command..."

George C. Wilson, *Supercarrier: An Inside Account of Life Aboard the World's Most Powerful Ship, the USS John F. Kennedy* (Berkeley Publishing Group, 1992). A veteran defense journalist of 50 years, Wilson takes the reader inside the armed forces for an unvarnished look at military affairs from the literal ground level up.

William L. Smallwood, *Warthog: Flying the A-10 in the Gulf War* (Brassey's, 1993). "Smallwood, who interviewed 143 of the pilots who flew the A-10 in the Gulf War, here presents an exhilarating, fact-packed narrative that conveys the emotional as well as the technical/tactical aspects of the Warthog effort during Desert Storm." The A-10 proved extremely effective in both Gulf wars and Afghanistan, while at the same time was among the most survivable per combat sortie.

Congress

Robert G. Kaiser, *So Damn Much Money: The Triumph of Lobbying and the Corrosion of American Government* (Vintage, 2010). "The life story of Washington lobbyist Gerald Cassidy is used to illuminate how Washington has changed over the past three decades in this bleak but informative book...The author also lays out a larger history of influence peddling in federal politics, stretching back to the Civil War era, and examines the evolution of today's permanent campaigns..."

Winslow T. Wheeler, *The Wastrels of Defense: How Congress Sabotages US Security* (U.S. Naval Institute Press, 2004). The highs and lows of a 30-year career on Capitol Hill to assess what makes Congress tick on national security issues.

Web Sites

YouTube Videos of John Boyd. At http://www.youtube.com/watch?v=U_fjaqAiOmc&feature=related, find a series of videos of John Boyd's presentations to audiences at the Air War College at Maxwell Air Force Base in the 1990s. Other video materials, of varying quality, about Boyd and his work are also there. Search the column on the right at the Web site.

Don Vandergriff's Web site. Vandergriff's Web site at http://www.donvandergriff.com/index.html, self described as "Whether it's in business, health care, law enforcement or national security, the 21st century is filled with volatility and uncertainty. How do we develop leaders who are adaptable, agile and able to help our organizations evolve in the face of an unpredictable environment?"

Defense and the National Interest contains the works of many military reformers, including those of John Boyd, Franklin C. Spinney, Chet Richards, William S. Lind and others. Find these extensive materials at http://dnipogo.org/, along with the text and associated materials of this handbook.

The *Straus Military Reform Project* at http://www.cdi.org/program/index.cfm?programid=37 or www.cdi.org/smrp contains the articles, commentaries and reports of multiple military reformers, including Winslow Wheeler, Pierre Sprey, Thomas Christie, Franklin C. Spinney, Col. Douglas Macgregor and others. This Web site is part of the Center for Defense Information network of Web pages at www.cdi.org. Both URLs above contain the text and associated materials of this handbook.

Project on Government Oversight, at http://www.pogo.org/, support multiple efforts associated with military reform and good government and describes itself as "an independent nonprofit that investigates and exposes corruption and other misconduct in order to achieve a more effective, accountable, open, and ethical federal government."